NEW RESEARCH ON FRAGILE X SYNDROME

NEW RESEARCH ON FRAGILE X SYNDROME

JEFFREY J. UPNER
EDITOR

Nova Science Publishers, Inc.
New York

LIBRARY OF CONGRESS CATALOGING-IN-PUBLICATION DATA
Available upon request

ISBN 13 978-1-60021-281-9
ISBN 10 1-60021-281-6

Published by Nova Science Publishers, Inc. ✢ New York

CONTENTS

PREFACE

Fragile X syndrome (FXS), is the most common cause of inherited mental impairment. This impairment can range from learning disabilities to more severe cognitive or intellectual disabilities. (Sometimes referred to as mental retardation.) FXS is the most common known cause of autism or "autistic-like" behaviors. Symptoms also can include characteristic physical and behavioral features and delays in speech and language development. This new book presents the latest research in this field.

A significant minority of children with fragile X syndrome (FXS) present with symptoms of autism; however, early identification of "true" autism cases is hampered by symptom overlap and instability of social and communicative features in individuals with FXS. Understanding the early emerging language profiles in FXS, specifically early emerging patterns of receptive and expressive language abilities, may provide a helpful clinical tool for identifying children with FXS who also have autism. Data from their studies in chapter 1, suggest that a young child with FXS who presents with significant impairments in receptive language abilities and depressed growth of skills in this area over time should be referred for a further evaluation for co-morbid autism. Implications for assessment practice and treatment will be discussed.

The present study in chapter 2, assessed cognitive functioning in sixty participants with fragile X syndrome (30 males and 30 females), ranging in age from 6 years to 34 years, divided into two separate groups distinguished by Chronological Age (under 16 years and over 16 years). Intellectual level ranged from 3 years to 13 years in affected males and from 3 years to 20 years in affected females. Performance was compared to typically developing participants matched on intellectual level and ranging in chronological age from 4 years to 14 years. A comprehensive battery of cognitive measures was employed with both groups, covering the domains of language, articulation, and spatial cognition (visual-perceptual, visuo-spatial and visuo-construction). Two important findings emerged: First, contrary to previous research, neither fragile X males nor females displayed a reduction in performance with increasing age across any of the cognitive measures. Indeed, performance on all tasks improved with verbal mental age (VMA) to the same degree in both fragile X and typically developing groups. Second, the authors found evidence of a gender-specific Fragile X 'signature' whereby males and females of similar verbal mental age display different cognitive profiles, with males especially vulnerable to impairments on skills that required a visuo-motor component. Together, these findings represent a unique contribution to the existing literature. The theoretical and practical implications will be discussed.

Fragile X Syndrome (FXS) is the most frequent cause of hereditary mental retardation. The neuropsychological manifestations of FXS, besides mental retardation (MR), also include a wide range of symptoms, such as behavioural manifestations and cognitive deficit. On the other hand, symptomatology of FXS is not limited to manifestations in individuals with the full gene mutation, but must also include the eventual consequences in bearers of premutation. The aim of this study in chapter 3, is to contribute to a better understanding of the syndrome with the following objectives: 1) assess the frequency of the most relevant clinical manifestations in FXS; 2) describe the cognitive behaviour phenotype of FXS in a sample of Spanish patients; 3) determine whether the phenotype found is specific to FXS, or whether it could also be related to MR or to Attention Deficit/Hyperactivity Disorder (ADH/D) inherent to the syndrome; and 4) provide information about the consequences of the premutation in cognitive and behavioural aspects. In order to fulfil these objectives, we carried out the following studies: Study 1 (106 patients): based on a questionnaire about medical and behaviour problems in FXS. Study 2 (30 FXS patients and two control groups: 30 MR and 320 ADH/D): a comparison between behaviours in FXS and behaviours found in patients with MR of other types and in patients with ADH/D. Study 3 (7 patients): neurocognitive manifestations found in premutated patients. Conclusions from the sample of patients with FXS, the clinical characteristics typical of the syndrome can be described, as well as their frequency. A clear prevalence of epilepsy (14%) can be observed, a figure lower than those recorded in most studies relative to this phenomenon. The comparison with the MR and ADH/D groups show that behavioural manifestations in FXS are inherent to the genetic mutation, as they cannot be explained on the base of MR and/or ADH/D, which is almost always present in FXS. The assessment of bearers of the premutation of FXS proves the presence of cognitive symptoms in premutated males. This information reinforces the genetic model according to which the clinical manifestations of FXS are distributed in a continuous way, from low-range premutation to high-range full mutation.

Fragile X syndrome is one of the most common hereditary forms of mental retardation, and has an estimated incidence of 1 in 4,000 males. The main characteristics of fragile X syndrome in males include mental retardation, macroorchidism, and peculiar face (elongated narrow face with large protruding ears and prominent jaw). In order to enlarge the number of variables useful in the preliminary evaluation of mentally retarded individuals, an analysis of data from medical records, physical examination, and anthropometric measurements have been undertaken in chapter 4. Forty two mentally retarded boys (age 4 to 20 years) with the characteristic features of fragile X syndrome, such as large protruding ears and elongated face were studied. Psychological examination including IQ and behavior disturbances was performed. Medical data, including birth weight were analyzed. Careful physical examination with special attention paid to minor malformations was performed. Anthropometric measurements were conducted in both groups. The measured variables included height, body weight, upper extremity length, lower extremity length, hand length, hand width, head length, head width, head breadth, total face height, bizygomatic diameter, bigonial diameter, nose length, nose width, ear height, ear width, inner canthal distance. Additionally, seven following indexes were calculated: cephalic, total face, mandibular-zygomatic, ear, nose, ratios of upper extremity length to stature high and lower extremity to stature high. Normative values for Polish children were obtained from published papers. Only in 16 boys methylated full mutations in the FMR1 gene were found, in 26 of them no recognizable cause of mental retardation was detected. The boys were divided into two groups: Fragile X and non-Fragile

X. High arched palate (44.44% v.s. 19.23%) and blue irises (83.33% v.s. 54.54%) were more often observed in FraX boys. The comparison between the average values of age adjusted anthropometric measures on all Fragile X and non-Fragile X boys showed significant differences with respect to height, head width, total face height and to the indexes of cephalic, ear and lower extremity length to stature. The greatest differences were found in ear width and weight. Two of the Fragile X boys, 10 and 11 years old brothers with marked obesity (+ 4,89 SD and + 4,6 SD) were initially suspected to have Prader-Willi syndrome.

Fragile X syndrome is characterized by mental retardation and a behavioral phenotype that includes stress-related abnormalities in behavior. These include gaze aversion, hyperactivity, hyper-arousal, hyper-responsivity to sensory stimuli, impulsivity, social anxiety, and social withdrawal. Recent results in chapter 5, indicate that the function of the hypothalamic-pituitary-adrenal axis, which is involved in the physiological and behavioral responses to stress, is altered in children with fragile X. Work in the authors' laboratory has begun to investigate the stress response in a mouse model of fragile X, the Fmr1-knockout, to determine the extent to which the hypothalamic-pituitary-adrenal axis is dysfunctional and to identify potential cellular targets for therapeutics. Using a restraint stress paradigm, they find that, similar to the human condition, Fmr1-knockouts have an exaggerated glucocorticoid response to stress indicating perturbed negative feedback in this system. In addition, stress-induced changes in the expression of the immediate early gene c-fos within hippocampus, neocortex, and the hypothalamic paraventricular nucleus, as well as expression of brain-derived neurotrophic factor within hippocampus, are strikingly altered in the fragile X mouse. These results, as well as other findings to be described in this chapter, demonstrate that the Fragile X mutants exhibit a greater response to stress than do wildtype mice, and support the hypothesis that the hypothalamic-pituitary-adrenal axis is dysregulated in fragile X syndrome.

Fragile X syndrome is the most common form of inherited mental retardation. Molecular basis of the syndrome is a CGG expansion in 5' untranslated region of the FMR1 gene. It is inherited as X linked dominant trait, with a reduced penetrance (80% for males and 30% for females). The general mechanism by which the unstable region is transmitted to the next generation has been established: full mutation expansion from premutated alleles is only acquired via maternal meiosis, with an increased risk for the larger premutation sized alleles. On the other hand, paternal transmission always remains in the premutated range, from both premutated and full-mutated fathers. Genetic counseling and prenatal diagnosis for this syndrome have substantiality improved using molecular testing. Nowadays and thanks to the new advances in assisted reproductive techniques such as donor oocytes, preimplantational diagnosis or prenatal diagnosis, the reproductive options for carrier females have considerably increased. The knowledge of the precise CGG repeat number in premutated carriers, allows us to perform an accurate genetic counseling.Moreover, new advances in fragile X syndrome have resulted in the description of pathologies associated to premutation carriers. Between 12-20 % of premutated women present premature ovarian failure (menopause before 40 years), while only 1% of general population presents it. On the other hand, one third of normal transmitting males and an unknown percentage of premutated women, present a new syndrome of tremor ataxia associated to fragile X premutation (FXTAS) after 60 years. These findings have modified genetic counseling for fragile X syndrome, as they should be taken into account.In this section, we analyze the genetic counseling given to 300 fragile X families through more than 10 years of experience. Who, when and where have to receive the information and how it has to transmitted to the rest of the family. The authors report the reproductive options chosen by families on basis of our 150 prenatal diagnoses experience. Finally, they also explain the impact received by the families when the information about premature ovarian failure and FXTAS syndrome is given.In summary, chapter 6, shows our experience in genetic counseling, prenatal diagnosis and associated pathologies for 300 fragile X families.

The Fragile X mental retardation protein FMRP is an RNA binding protein that associates with a host of mRNAs and with other cellular RNA binding proteins to form messenger ribonucleoprotein particles (mRNPs). These mRNPs, in turn, associate with ribosomes and it has been shown that FMRP selectively represses the translation of some of the messages to which it binds. In addition, FMRP is a component of a variety of supramolecular RNA/protein granules including: the RNA induced silencing complex (RISC) complex, stress granules, transport granules and mRNA granules. In chapter 7, the authors have recently identified a unique and smaller cytoplasmic FMRP granule. The nature of this granule is not fully characterized, yet it is known that its presence increases when cellular methylation is blocked. Here, they demonstrate that small cytoplasmic FMRP granules are a common feature of cultured cell lines. They also show by *in situ* hybridization to an $Alexa_{488}$-oligo dT_{18} probe that this granule is devoid of poly A^+ RNA. Furthermore, in cells treated with the methylation inhibitor AdOx prior to the induction of stress granules many FMRP-granules are physically associated with, albeit distinct from, stress granules. These and other circumstantial data suggest that the small cytoplasmic FMRP granules found in cultured cells might be mRNA processing bodies (P-bodies). Endogenous FMRP is recruited into abortive pre-initiation complexes (48S*) called stress granules following heat shock, treatment with sodium arsenite and mechanical injury. These granules are the major repository for FMRP following oxidative stress and we show that while FMRP and the other Fragile X-related proteins (FXRPs) form stress granules, not all proteins do so. This indicates that although heterogeneous in size, stress granules have a limited composition. We also report that the neurites of arsenite-treated PC12 cells harbor a heterogeneous population of FMRP granules. Some co-localize with the T cell internal antigen 1 (TIA-1), an RNA binding protein that defines classic stress granules. Some co-localize with FXR1P or FXR2P. However, in each instance the co-localization was incomplete. This is in contrast to the arsenite-induced FMRP granule populations in the cell soma and in the growth cones, which by-and-large uniformly contain all four proteins. Our data support a model in which sodium arsenite treatment provides a snapshot of granule trafficking by marking potentially translational mRNA granules from those that have delivered their messages and are transiting back to the nucleus for others. Thus, FMRP lies at the confluence of many types of granules and the underlying biology of these granules may provide the molecular basis for the wide-ranging phenotype seen in fragile X syndrome.

In: New Research on Fragile X Syndrome
Editor: Jeffrey J. Upner, pp. 1-21
ISBN: 1-60021-281-6

Chapter 1

Language Profiles in Young Children with Fragile X Syndrome and Autism Symptoms

Amy Philofsky and Susan L. Hepburn
The University of Colorado at Denver and Health Sciences Center, USA

Abstract

A significant minority of children with fragile X syndrome (FXS) present with symptoms of autism; however, early identification of "true" autism cases is hampered by symptom overlap and instability of social and communicative features in individuals with FXS. Understanding the early emerging language profiles in FXS, specifically early emerging patterns of receptive and expressive language abilities, may provide a helpful clinical tool for identifying children with FXS who also have autism. Data from our studies suggest that a young child with FXS who presents with significant impairments in receptive language abilities and depressed growth of skills in this area over time should be referred for a further evaluation for co-morbid autism. Implications for assessment practice and treatment will be discussed.

The Study of Behavioral Phenotypes and Language Symptoms

Nearly 20 years ago, Burack, Hodapp, & Zigler (1988) proposed that individuals with mental retardation be studied and grouped according to specific etiologies, or pathways to a specific syndrome. This concept was an outgrowth of Zigler's (1969) "two-group approach" (i.e., organic vs. familial) to the study of mental retardation. Etiology-based groupings, these authors contended, would provide for more exact research and a better understanding of mental retardation than grouping individuals based upon level of functioning (Burack et al., 1988). Behavioral phenotype research exploring the characteristics common and unique to different disorders subsequently emerged in the field of developmental disabilities in the 1980's and 1990's.

Language symptoms are often observable features associated with the behavioral phenotypes of different mental retardation syndromes (Rice, Warren, & Betz, 2005). Cross-syndrome comparisons of specific language features may provide insight into the shared language symptomology of developmental disorders, which is important for several reasons. Not only does this type of science help to clarify the characteristic symptoms of each disorder, but it also helps to clarify different ways in which human language development can be vulnerable (Rice et al., 2005). Additionally, critical pieces of the definition of behavioral phenotypes, including subgroups within behavioral phenotypes, may be elucidated (Rice et al., 2005). Finally, awareness of potential language symptoms for some developmental disorders may prove useful for signaling clinicians of the need for further evaluation into possible co-morbid diagnoses (Philofsky, Hepburn, Hayes, Rogers, & Hagerman, 2004).

Such research has particular implications for two disorders that share language symptomology among their hallmark features, such as fragile X syndrome (FXS) and autism. While FXS remains among the least understood of disorders with known behavioral phenotypes in terms of language symptoms (Rice et al., 2005), the language symptoms of autism have been much more heavily studied. Autism occurs in approximately one-quarter to one-third of individuals with FXS, and FXS remains one of few known genetic causes of autism (Bailey, Hatton, Skinner, & Mesibov, 2001; Cohen, 1995; Hagerman, Ono, & Hagerman, 2005; Turk & Graham, 1997). Knowledge of symptom overlap between these two particular disorders has the potential to provide some insight into causal mechanisms. In the particular case of these two disorders, overlapping language features might suggest similar etiologies. Likely the same affected pathways in co-morbid FXS and autism have similarities to other subgroups of autism (Hagerman et al., 2005). Thus, elucidating the language and cognitive profiles of children with FXS, autism, as well as co-morbid autism and FXS provides an important area for inquiry.

Overview of Fragile X Syndrome

FXS, the most common heritable form of mental retardation, is associated with physical features, language delays, and a characteristic behavioral phenotype (Hagerman, 1999). Caused by an abnormality located near the long arm of the X chromosome at Xq27.3, this abnormality involves insufficient production of a protein called Fragile X mental retardation (FMRP), a deficiency that has been associated with abnormalities in brain development (Hagerman, 1999). The associated abnormalities in production of FMRP further lead to expansion of the CGG nucleotide sequence on the X chromosome, such that expansion beyond 200 repeats results in the syndrome (repeats of up to 50 are considered normal, while repeats between 50 and 200 are considered a premutation of the syndrome; Reiss & Dant, 2003). Diagnosis is made via FISH (Fluorescent In Situ Hybrid) testing (Hagerman, 1999). Prevalence estimates range between 1:3200 to 4000 in males and 1:6000 in females (Sherman, 2002) and the syndrome is represented equally across ethnic groups (Reiss & Dant, 2003). Males with FXS are generally more adversely affected than females, due to a female's biological compensation of an unaffected X chromosome. Researchers are investigating the relationships between the extent of functional impairment with genetic factors, such as the percentage of protein present and the individual's methylation status (Hagerman, 1996). For a thorough review of the molecular genetics associated with the development of FXS, please see Hagerman (1999).

Over the past few decades a distinct physical and behavioral phenotype has been emerging for FXS. In terms of associated physical features, macroorhidism (i.e., enlarged testicles), an elongated face, with large, prominent ears and jaw, increased head circumference, and heart abnormalities have all been reported (Hagerman, 1999). In terms of the behavioral phenotype associated with FXS, three major areas have been identified as characteristically affected: cognition, behavior, and language.

In terms of cognition, boys with fragile X syndrome tend to demonstrate mild to moderate mental retardation, while girls generally demonstrate milder symptoms (Hagerman, 1999). Several investigators have suggested that cognition follows an abnormal trajectory in FXS, such that beginning as early as age 5, children often begin to show a decrease in IQ as a function of a slowing and/or plateauing of cognitive development (as reviewed in Reiss & Dant, 2003). Longitudinal analysis considering the cognitive trajectories of 46 young boys with FXS (ages 24-72 months) suggested that, on average, the growth rate was approximately half of typical development though significant variability was noted in their sample (Bailey, Hatton, & Skinner, 1998). By contrast, Freund, Peebles, Aylward, & Reiss (1995) found that 44% of their preschool-aged sample of 18 boys with FXS tested in the borderline to average IQ range.

Additionally, rather than demonstrating overall global weaknesses, as is the case with most individuals with non-specific forms of mental retardation, individuals with FXS often demonstrate a characteristic profile of relative cognitive strengths and weaknesses. Specific areas of strength have been identified to include: simultaneous processing (i.e., understanding and organizing information allowing for the "whole" to bring meaning to the "parts;" Kemper, Hagerman, & Altshul-Stark; 1998) and imitation abilities (Rogers, Hepburn, Stackhouse, and Wehner, 2003). Areas of particular difficulty identified have included: sequential processing (understanding and organizing information allowing for the "parts" to bring meaning to the "whole;" Freund & Reiss, 1991), visual-motor coordination (Freund & Reiss, 1991) and short-term visuo-spatial memory (Dykens, Hodapp, & Leckman, 1987).

Turning to behavior, the following areas have been identified as particularly impacted within the FXS behavioral phenotype: impulsivity, attention, hyperarousal, and anxiety (Cornish et al., 2004). Co-morbid diagnoses of attention-deficit hyperactivity disorder have been reported to occur in as many as 73% of school-aged boys with FXS (Hagerman, 1987) with its presentation seen in nearly twice as many boys with FXS compared to children with non-specific forms of mental retardation (Borghgraef, Fryns, Dielkins, Pyck, & Van Den Berghe, 1987). Furthermore, difficulties with anxiety are frequently reported to be associated with difficulties engaging in social interactions, including social avoidance and withdrawal (Freund et al., 1995; Hessl et al., 2002). Finally, hyperarousal may be evidenced in the following characteristic behaviors: poor eye contact, tactile defensiveness, hyperactivity, handflapping, nailbiting, and tantrums (Reiss & Dant, 2003).

Language delays, particularly in the area of expressive language, are typical (Rice et al., 2005; Roberts, Mirrett, & Burchinal, 2001), though receptive language has been reported to be commensurate with mental age by the time boys reach adolescence and young adulthood (Abbeduto et al., 2003). A relative strength in receptive language compared to expressive language has been suggested to contribute to the FXS communicative phenotype (Roberts et al., 2001). Also notable, a number of atypicalities are observable in the expressive language of individuals with FXS including: increased rate and impulsivity of speech production (Borghgraef et al., 1987; Hanson, Jackson, & Hagerman, 1986; Wolf-Schein et al., 1987), dysfluency (Hanson, et al., 1986; Newell, Sanborn, and Hagerman, 1983), repetitive language (Borghgraef et al., 1987; Sudhalter, Cohen, Silverman, & Wolf-Schein, 1990), and articulation and sound sequencing difficulties (Hanson et al., 1986; Newell et al., 1983; Paul, Cohen, Breg, Watson, & Herman, 1984). Furthermore, descriptive explanations have evolved to describe some notable differences in the speech of individuals with FXS, including such terms as, "jocular," "staccato," "litany-like phraseology," and "sing-songy" (Dykens, Hodapp, & Leckman, 1994; Turner, Daniel, & Frost, 1980).

Overview of Autism

Autism is a behaviorally-diagnosed, neurobiological disorder that creates impairments in three core areas of functioning: communication, social relatedness, and restricted or repetitive interests (APA, 2003). Difficulties with basic social relatedness and reciprocity comprise the hallmark feature of an autism spectrum diagnosis (Volkmar et al., 1994). Currently the behavioral phenotype for autism is heterogeneous, such that children sharing an autism spectrum diagnosis vary significantly in level of functioning despite difficulties in common across the three core areas. Researchers continue to struggle in identifying meaningful subgroups of children sharing this diagnosis (Szatmari, 2000). Children with comorbid autism and FXS comprise a meaningful subgroup of persons on the autism spectrum. This subgroup of children usually presents with severe mental retardation and a deviant pattern of language development related to the differences in communication development that are observed in all but the very earliest stages of language development (Bailey et al., 2001; Lord, Risi, & Pickles, 1994; Rice et al., 2005).

Although previously considered a rare disorder, autism rates have risen alarmingly to approximately 1:300 to 1:800 children, placing it among the more common disorders of childhood (APA, 2003; Fombonne, 2005; Yeargin-Allsopp, Rice, Karapurkar, Doernberg, & Murphy, 2003). Mental retardation is highly variable and sometimes co-morbid with autism, occurring in roughly 25-80% of individuals (Chakrabarti & Fombonne, 2001; Fombonne, 1999), with most children exhibiting co-morbid autism and fragile X syndrome falling into the severe mental retardation range (Bailey et al., 2001). With fragile X syndrome an exception, the exact cause remains unknown- although it has been speculated that autism results from a combination of both genetic and environmental influences upon fetal development (Bailey, Phillips, & Rutter, 1996).

With variability underscored in the autism behavioral phenotype, precision is compromised in limiting the disorder to one single behavioral phenotype. However, because autism is behaviorally diagnosed, the behaviors of individuals with autism have been intensely studied and distilled into specifics that are necessary for making such a diagnosis (APA, 2003). In autism, the cognitive-linguistic profile develops in splintered fashion, with non-verbal functioning typically emerging as an area of strength, relative to weaknesses in verbal abilities (APA, 2003), although Siegel, Minshew, and Goldstein (1996) found that only 40% of their well-characterized high-functioning (i.e., IQ >70) sample of 45 children with autism fit this profile, while over half of their sample (i.e., 58%) demonstrated the opposite profile (i.e., VIQ>PIQ). However, a relative nonverbal strength is especially evident in the approximate one-fifth of children with autism who remain nonverbal (Lord et al., 2004).

Language development is usually delayed in autism (APA, 2003). While vocabulary and grammar acquisition (i.e., structural language) in autism spectrum disorders range from normal to delayed, difficulties with understanding and appropriately using language (i.e., pragmatics) are universal, including difficulties with the use of nonverbal forms of communication (Dawson et al., 2002; Happe & Frith, 1996; Mundy, Sigman, Ungerer, & Sherman, 1986; Tager-Flusberg, 2000). A number of unusual expressive language features have been associated with autism including: echolalia; use of neologisms (i.e., meaningless, idiosyncratic words); and pronoun reversals (Lee, Hobson, & Chiat, 1994; Tager-Flusberg, 2000; Volden & Lord, 1991). Also, children with autism spectrum disorders tend to be literal and concrete and have difficulty with abstract ideas in language and social reasoning, including difficulties with both social-perceptual and cognitive Theory of Mind tasks, or the ability to interpret others' thoughts based upon social-emotional cues and/or inference (Baron-Cohen, 1989; Minshew, Meyer, & Goldstein, 2002). The combination of these factors contributes to difficulties for children with autism spectrum disorders in interpreting environmental cues (Marcus, Garfinkle, & Woolery, 2001; MDOE, 1998).

Co-morbid Fragile X Syndrome and Autism

Differential diagnosis of co-morbid autism and FXS can be very complex, with approximately one-quarter to one-third of children with FXS also meeting behavioral criteria for autism (Bailey et al., 1998; Bailey, Roberts, et al., 2004; Hagerman et al., 2005; Philofsky et al., 2004; Turk & Graham, 1997). To date, no large epidemiological studies have been conducted to address the prevalence of these co-morbid conditions. Smaller studies have reported a comorbidity rate of autism and FXS from as low as 6% (Brown et al., 1986) to as high as 40% (Rogers, Wehner, & Hagerman, 2001).

Diagnostic differentiation is complicated by the behavioral overlap observed between the phenotypes of FXS and autism. Specific examples of these behaviors include reactions to environmental stimuli, repetitive speech, self-injury, hand stereotypies, handbiting, spinning, and inappropriate and repetitive use of objects (Fisch, Simensen, & Schroer, 2002; Levitas, Braden, Van Norman, Hagerman, and McBogg, 1983; Turk & Graham, 1997). However, these overlapping behaviors tend to fall under only one of the three necessary behavioral domains that constitute an autism diagnosis: restricted and/or repetitive interests.

Rather, the communication and social characteristics (i.e., the other two domains necessary for an autism diagnosis) tend to differentiate the two syndromes when they do not co-morbidly present (Kau et al., 2004; Kaufmann et al., 2004). More specifically, persons with FXS may avoid eye contact with others and may withdraw from social interaction; however, core social relatedness – particularly with familiar people in familiar situations – is intact. The social interaction deficits noted in FXS have been hypothesized to be related to anxiety and hyperarousal, which is not the case in autism.

Thus, the research suggests that there are two distinct phenotypes within FXS, based upon the absence or presence of autism (Cohen, 1995; Kau et al., 2004; Kaufmann et al., 2004; Philofsky et al., 2004; Rogers et al., 2001). Beyond presenting with common repetitive and/or restricted interests, children with FXS who also have an autism spectrum disorder, additionally present with persistent qualitative impairments in both social and communicative symptoms (Kau et al., 2004). Like children on the autism spectrum, the social behavior of children diagnosed with co-morbid FXS and autism occur on a continuum ranging from mild to severe impairment (Kauffmann et al., 2004). Further, while findings from Kau and her colleagues (2004) suggested that co-morbid autism stemmed from more significant communication impairments, in a companion paper Kauffmann and his colleagues (2004) clarified that a co-morbid diagnosis of autism in FXS was most significantly related to core social deficits, as is the case with autism spectrum disorders (Hobson, 1993; Kau et al., 2004; Kauffmann et al., 2004; Sigman, 1994). For a child with FXS to meet social criteria for autism, difficulties in integrating eye gaze with functional communication, shared attention and affect should be observable, and social withdrawal should not improve as familiarity increases.

Children who present with co-morbid diagnoses of autism and FXS typically present with poorer performance on cognitive, communication, social, adaptive, and behavior variables than children with either diagnosis alone (Bailey et al., 2001; Cohen, 1995; Philofsky et al., 2004; Kau et al., 2004; Kauffmann et al., 2004; Rogers et al., 2001, 2003). Specifically, more severe mental retardation, lower receptive and expressive language skills, poorer imitation skills, poorer adaptive skills, worse social deficits, and more severe behavior problems are seen in children with co-morbid autism and FXS than children who present with only a diagnosis of either autism or FXS alone (Bailey et al., 1998; Kau et al., 2004; Kauffmann et al., 2004; Philofsky et al., 2004; Rogers et al., 2001, 2003). While it is becoming evident that children with co-morbid autism and FXS tend to be more developmentally impaired overall than most children with either disorder alone, a small but growing body of research has suggested that the behavioral phenotype of co-morbid autism and FXS more closely resembles autism than FXS (Kau et al., 2004; Kauffmann et al., 2004; Philofsky et al., 2004; Rogers et al., 2001, 2003).

THE LANGUAGE PROFILE IN FXS

Language profiles associated with x-linked retardation syndromes have demonstrated a trend towards relatively stronger receptive than expressive language abilities (Newell et al., 1983); however, there have been only a few studies considering the specific language profiles of individuals with FXS. In spite of overall language delays, better receptive than expressive language was noted for three boys (ages 13:9 [years:months], 10:6, and 10:0) with FXS (Paul et al., 1984). In 1987, Paul and colleagues provided preliminary evidence for rapid growth of receptive language ability during early childhood in males with FXS. Roberts and colleagues (2001) reported that receptive language skills were superior to expressive language skills in 39 young males (ranging in age from 20 – 86 months) with FXS and concluded that this splintered pattern in the development of language skills may be relevant for the construction of a communication phenotype of young boys with FXS (Roberts et al., 2003). Furthermore, by young adulthood, receptive language has been reported to be mental-age appropriate in boys with FXS who do not meet criteria for an autism spectrum disorder (Abbeduto et al., 2003).

Work on a longitudinal study, following 2-year old children with FXS over time, replicated this relative strength in the receptive language abilities of even younger children with FXS (mean age 34 months) who did not carry an additional co-morbid diagnosis of autism (Philofsky et al., 2004). Thus with respect to the language profile, individuals with FXS tend to show a relative strength in receptive language (equivalent to nonverbal cognition) that can be identified in children as young as 34 months, with marked weaknesses in the development of expressive language (Abbeduto et al., 2003; Paul et al., 1984, 1987; Philofsky et al., 2004; Roberts et al., 2001).

THE LANGUAGE PROFILE IN AUTISM

In contrast to children with FXS, young children with autism tend to have significant receptive language delays, which have been compared to those observed in developmental receptive dysphasia (Bartak, Rutter, & Cox, 1977). In fact, the severity of the receptive language impairments observed in autism has sparked debate over the possible coexistence of developmental language disorder and autism (Bartak, Rutter, & Cox, 1975; Konstantareas & Beitchman, 1996). Tager-Flusberg (2000) noted that the primary social deficits of autism fail to set the stage for the development of language and communication in early childhood; with limited social awareness in early childhood, interactions with others on any level, especially language, are deferred (Tager-Flusberg, 2000).

Despite notable weaknesses in receptive language abilities in children with autism, relatively even receptive and expressive language profiles have been reported for children with autism (Fein et al., 1996; Jarrold, Boucher, & Russell, 1997; Kjelgaard & Tager-Flusberg, 2001). Fein and colleagues (1996) reported on a large-scale study involving 176 preschool children with autism. They concluded that children with autism demonstrated a relatively even language profile, with slightly stronger expressive than receptive abilities. Jarrold, Boucher, & Russell (1997) also found that the language profiles for 120 children on the autism spectrum ranging in ages from 5 years 6 months to 19 year 7 months, suggested relatively even development, with respect to receptive and expressive language abilities. Finally, Kjelgaard & Tager-Flusberg (2001) examined the language profiles of 89 children with autism between the ages of 4 and 14-years-old and found that verbal children with autism tended to show relatively even performances in receptive and expressive language domains. This finding of a relatively even receptive and expressive language profile in autism has been replicated in children as young as 34 months (Philofsky et al., 2004).

LANGUAGE PROFILES: CHILDREN WITH COMORBID FXS AND AUTISM

Several of the language characteristics observed in persons with FXS, particularly in the area of expressive language, are also observed in individuals with autism (Newell et al., 1983). Wolf-Schein and colleagues (1987) identified the following speech-language characteristics that are observed in both disorders: jargon, perseveration, echolalia, inappropriate and tangential language, and talking to self. In light of this overlapping symptomology, expressive language features may not provide much specificity as to the distinguishable behavioral phenotypes of autism and FXS.

Our lab recently published results examining this hypothesis in young children relating to the linguistic profile of children with co-morbid FXS and autism (Philofsky et al., 2004). With respect to early language profiles, would these children look "more like FXS," demonstrating a receptive language advantage in early childhood? Or would the language profile of children with co-morbid FXS and autism look "more like autism," demonstrating a relatively even linguistic profile in early childhood?

Using state-of-the-art tools for the diagnosis of autism symptoms (i.e., the ADOS-G and ADI-R), we included children with autism (n = 18) and children with fragile X syndrome (n = 18), and assessed every child with FXS for a co-morbid diagnosis of autism, resulting in three groups of children: children with FXS (n = 10); children with autism (n = 18); children with co-morbid FXS and autism (n = 8). Examination of the autism symptomology in the autism group versus the co-morbid FXS and autism group revealed no observable group differences.

Table 1. Research articles examining the linguistic profiles in autism, FXS and co-morbid autism and FXS

Article	Study Group	Participants	Profile Findings
Madison, George, and Moeschler, 1986	FXS	12 family members with FXS aged 4-63-years-old	EL >RL*
Paul et al., 1984	FXS	3 boys aged 10-13-years-old	RL>EL
Paul et al., 1987	FXS	12 adult males	RL>EL
Roberts et al., 2003	FXS	39 boys aged 20-86 months	RL>EL
Jarrold, Boucher, & Russell,1997	autism	120 children and adolescents aged 5 years, 6 months to 19 years, 7 months	RL=EL
Kjelgaard & Tager-Flusberg, 2001	autism	89 verbal children aged 4-14-years-old	RL=EL
Rapin, 1996	autism	176 children aged 36 to 83 months	RL=EL
Philofsky et al., 2004	autism, FXS, and FXS w/ autism	36 children aged 22-45 months; 18 with autism, 10 children with FXS, and 8 children with FXS and autism	FXS: RL>EL; autism: RL=EL; Autism/FXS: RL=EL, severely depressed relative to both other groups

*RL= Receptive Language; EL =Expressive Language

Lacking early core social deficits were evident in the children with autism and co-morbid FXS and autism, but not in the children with FXS. Examination of the linguistic profiles (i.e., receptive vs. expressive language) in these three groups supported our hypothesis: Young children with FXS, who do not present with a relative strength in receptive compared to expressive language functioning, were more likely to present with co-morbid autism. As it turned out, our young co-morbid FXS and autism group looked "more like autism" than FXS, in terms of early receptive language functioning by not showing the relative receptive language strength seen in young children with FXS without autism (Philofsky et al., 2004). See Table 1 for a summary of articles describing the language profiles in children with autism, FXS, and co-morbid autism and FXS.

However, like other reports of children with co-morbid autism and FXS, our cohort were lower functioning than children in either of the other two syndrome groups. Of the 8 children in the fragile X (with autism) group, 6 (86%) presented with overall mental ages of 18 months or less, which certainly affects the pattern of social, communicative, and behavioral functioning observed in these children. Only 3 children with fragile X without autism (25%) presented with mental ages below 18 months Some researchers have suggested that the comorbidity of autism symptoms in persons with fragile X syndrome is related to cognitive factors (rather than to core social deficits), such that developmentally young and/or low-functioning children with fragile X are more likely to present with autism symptoms (Bailey et al., 1998; Borghgraef et al., 1987). The directionality of the effects observed in our study remained unknown: did having autism affect overall development or did overall development create a behavioral picture that looked like autism?

Additional Autism, FXS, and Co-morbid FXS/Autism Comparisons of Early Language Abilities

Additional data have been added for this chapter, expanding upon our earlier study and continues to include three groups of children qualified for autism diagnoses using the same methods described earlier in the chapter: 1) children with an autism spectrum diagnosis (ASD; $n = 36$) 2) children with a FXS diagnosis without co-morbid autism (FXS; $n = 16$) and 3) children with FXS and co-morbid autism (FXS/ASD; $n = 6$). Further, an additional longitudinal sample will be described consisting of a subset of the same sample seen approximately two years earlier at Time 1, however because of the small sizes of the Time 2 FXS groups, no inferential analyses will be performed. However, the data from Time 2 will be presented in Figure 1 (in addition to Time 1 data) and discussed descriptively, though our interpretations should be viewed with caution as a function of our small sample sizes in the FXS groups, in particular, at that time point. Attrition of the longitudinal sample, particularly in the FXS groups, was significant as a result of prohibitive travel expenses ($n = 9$), some families declining participation for unknown reasons ($n = 2$), and one family declining participation as a result of a heavy educational and therapy schedule ($n = 1$). While many of the children used for analyses for the first time point overlap with the sample from our earlier study (Philofsky et al., 2004), some additional children have been added. Please see Table 2 for the sample characteristics of the children depicted in Figure 1.

Table 2. Participant characteristics for the sample of children depicted in Figure 1.

	Time 1 ASD	Time 2 ASD	Time 1 FXS	Time 2 FXS	Time 1 FXS/ASD	Time 2 FXS/ASD
N	36	32	16	8	6	2
Diagnosis	33 autism; 3 PDD-NOS per ADOS and ADI-R	30 autism; 2 PDD-NOS per ADOS and ADI-R	Medical diagnosis of FXS	Medical diagnosis of FXS	Medical diagnosis of FXS and autism per ADOS and ADI-R	Medical diagnosis of FXS and autism per ADOS and ADI-R
CA (mos.)						
M (SD)	34.4 (4.1)	57.2 (5.0)	34.6 (8.5)	59.8 (8.6)	37.2 (6.2)	68 (7.1)
Range	26-46 mos.	48-72 mos.	23-50 mos.	52-74 mos.	27-45 mos.	63-73 mos.
NVMA			*n* = 15[1]	*n* = 7[2]		*n* = 1[3]
M (SD)	24.0(10.7)	39.0 (11.5)	22.4 (5.4)	32 (7.9)	17.8 (3.2)	27
VMA						
M (SD)	15.8 (7.2)	30.1 (14.5)	21.9 (10.0)	29.7 (7.8)	10.5 (3.0)	14.2 (4.6)
Gender						
Male (X%)	81%	81%	88%	100%	100%	100%

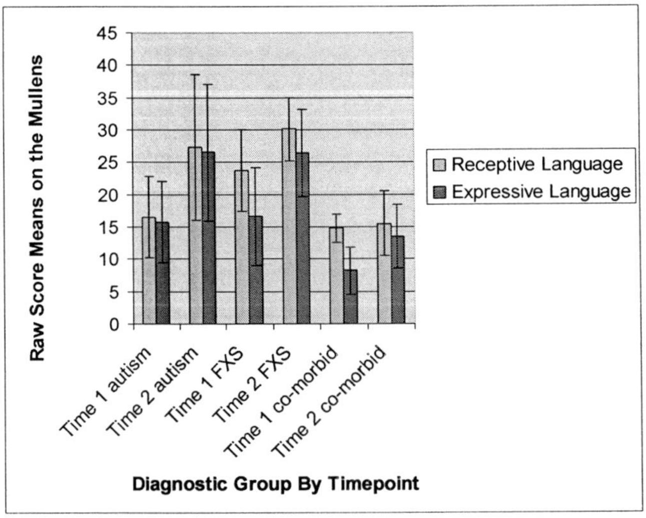

Figure 1. Average receptive and expressive language raw scores on the Mullens Scale of Early Learning in children with autism, FXS, and co-morbid autism and FXS at Time 1 (N = 58, 1-3-year-olds) and again at Time 2 (N = 42, 4-6-year-olds).

[1] NVMA could not be calculated for one child in this cell
[2] VMA could not be calculated for one child in this cell
[2] NVMA could not be calculated for one child in this cell

Due to the small number of participants in the FXS subgroups seen at Time 1, we conducted nonparametric analyses, specifically using the Kruskal-Wallis Test. If a mean difference was detected, pair-wise nonparametric testing was conducted in order to determine between which groups the differences lie. At Time 1, the children did not differ by chronological age, H (2, 58) = 1.7, p = .42, and none of the groups significantly differed in their mean nonverbal mental age (NVMA), although the power in the co-morbid autism/FXS group was limited by an n of only 6 and visual examination suggested somewhat lower functioning than the other two groups. Further, a Kruskal-Wallis Test was approaching significance, H (2, 57) = 5.7, p = .06. Pair-wise comparisons suggested significantly better NVMA for both the ASD and FXS groups than for the FXS/ASD group, H (1, 42) = 5.5, $p <$ .05 and H (1, 21) = 4.0, $p <$ .05, respectively. The FXS and ASD groups were not significantly different on NVMA AT Time 1. Mean differences were also detected in mean verbal mental age (VMA), H (2, 58) = 9.6, $p <$ .01, with the FXS group having significantly better VMA's at Time 1 than the ASD or FXS/ASD groups, H (1, 52) = 4.5, $p <$ .05 and H (1, 22) = 7.5, $p <$ .01, respectively. At Time 2, inferential tests could not be computed due to limited power in the FXS groups, although visual inspection suggested that all three groups look similar in terms of CA and NVMA. While the ASD and FXS groups look similar in terms of VMA, the FXS/ASD children appear worse on this variable.

Based on the data visually depicted in Figure 1, for children seen at Time 1, Kruskal-Wallis Tests were performed to determine if differences are occurring between group means on any of the three following variables: overall receptive language raw scores on the Mullens Scales of Early Learning (MSEL), overall expressive language raw scores on the MSEL, and the absolute value of the overall difference in mean raw scores of receptive and expressive language on the MSEL (absolute split raw score). While the Between Group factor is diagnosis (i.e., autism, FXS, or co-morbid FXS/autism), the Within Groups factors are the three language variables just listed from the MSEL. Raw scores were selected in order to avoid floor effects for the children functioning at the lowest levels on these variables on the MSEL (Mervis & Klein-Tasman, 2004).

Additionally at Time 1, the data reflected that there were mean group differences in overall mean receptive language raw scores, H (2, 58) = 13.1, $p <$ .01. While there were no differences between ASD and FXS/ASD on the receptive language raw score of the MSEL, the FXS group was significantly better than both the ASD group, H (1, 52) = 11.5, $p <$ .01 and FXS/ASD group, H (1, 22) = 7.3, $p <$ .01. Whereas there were significant mean differences in mean expressive domain raw scores, H (2, 58) = 7.4, $p <$ .05, the differences were occurring between the FXS/ASD and both FXS, H (1, 22) = 4.9, $p <$ .05, as well as the ASD group, H (1, 42) = 7.4, $p <$ .01, with no mean differences on expressive language raw score between the FXS and ASD groups. We also examined the size of the discrepancy between receptive and expressive abilities by diagnostic group by taking the absolute values of the difference between expressive language raw score and receptive language raw scores on the MSEL (split score). Results suggested significant group differences, H (2, 58) = 13.9, p $<$ .01. Specifically, ASD was different than both FXS and FXS/ASD, H (1, 52) = 8.4, $p <$ .01 and H (1, 42) = 8.6, $p <$ .01, respectively. The FXS and FXS/ASD groups were not statistically different from one another on the split score. These results reflect that while autism does not evidence a significant split score at Time 1 between receptive and expressive language raw skill development, both FXS groups do.

At Time 2, visual inspection of the data suggests that both ASD and FXS groups gain language skills and appear to be functioning at relatively similar levels in both receptive and expressive language, though children with FXS may continue to evidence slightly better receptive language abilities. The FXS/ASD children appear to gain skills in the area of expressive language, with fewer gains in receptive language and continue to lag behind both other groups in both receptive and expressive language abilities. The split in both FXS groups appears to decrease as expressive language skills increase, relative to receptive language skills over time, while the children in the ASD group continue to evidence relatively even overall receptive and expressive language development.

Overall, these data should be interpreted cautiously due to our small samples, however our data do suggest that 2-3 year old children with autism present with delays in both receptive and expressive language functioning, with few children evidencing a relative strength in either area. Notably, over time, many children with autism make significant progress in both receptive and expressive ability (as do children with FXS without autism). While children with FXS and autism more closely resemble the autism group, in terms of overall receptive language development at Time 1, these children tend to more closely resemble the FXS group when considering the relative differences between receptive and expressive language ability raw scores at the same age. At Time 2, however, the FXS and autism groups look more similar to one another as both groups grow in overall skills and the gap between receptive and expressive raw scores diminishes in the FXS group. Additionally, the FXS/ASD group may demonstrate the lowest overall language functioning at Time 2 and make some improvement in expressive language but little improvement over time in the receptive language domain, in particular (see Figure 1).

One area of potential distinction between FXS and autism concerns early receptive language abilities. While children with FXS tend to show a relative strength in receptive language (compared to expressive language), children with autism do not show this relative strength; rather, their language profiles tend to develop more depressed and evenly (Philofsky et al., 2004). Additionally, growth in receptive language abilities over time is the most suppressed in children with co-morbid autism and FXS compared to both young children with only autism and young children with only FXS (See Figure 1).

In summary, although our sample sizes are small (particularly at Time 2) and require cautious interpretation, for children with FXS without comorbid autism, the relative strength in receptive language continues; however, gains are also noted in expressive language, such that the gap between the two aspects of language narrows somewhat. For children in our study who presented with FXS and met extensive criteria for autism at age 2-3, marked weaknesses are noted in receptive and expressive abilities, with a reduction in the split between the two abilities over time as a function of increased expressive skill acquisition and a significantly limited growth in receptive language abilities over time. Finally, children with autism make considerable growth in both receptive and expressive language abilities and continue to evidence fairly even development across the two domains over time (See Figure 1).

Continuity of Autism Diagnoses in Young Children with FXS

Longitudinal analyses of our sample of children with FXS were necessary to clarify whether or not our co-morbid cohort actually presented with a true co-morbid diagnosis of autism- or were significantly developmentally delayed enough between the ages of 2 and 3 to create the false early illusion of autism. Although our FXS group experienced significant attrition (i.e., 7 children)[4], 11 children from our original study were seen again and re-evaluated for a co-morbid diagnosis of autism using state-of-the-art diagnostic measures (i.e., the ADI-R, ADOS, and DSM-IV checklist) at a subsequent time point (i.e., between the ages of 4 and 6). A Mann-Whitney test comparing the mean ranks of the Time 1 Early Learning Composites from the Mullens Scales of Early Learning (MSEL) of the 7 children who were only seen at Time 1 to the 11 children who were seen at Time 2 suggested that there were no significant differences in overall developmental functioning between these two groups (U = 28.5, ns), supporting the idea that the children who were seen longitudinally were not developmentally different than the children who were only able to be seen at the first time point.

Results suggested that 2 of the 11 children (18%) met criteria on both the ADI-R and ADOS-G for a co-morbid diagnosis of autism at both time points. None of the children met criteria at one (but not the other) time point on both measures. Therefore, autism symptomology in our sample was fairly stable across two years of development. Further, ADOS-G scores were strongly correlated with cognitive abilities, as measured by the nonverbal mental age of the Mullens Scales of Early Learning, at the first timepoint, such that as symptoms of autism increased cognitive functioning decreased ($r(9) = -.83, p < .001$). Note that non-verbal mental age scores were not available for two of the 11 children at the first timepoint. However, this relationship did not hold for the same group of children at the second timepoint ($r(11) = .30$, *ns*), such that by age 5, cognitive functioning and autism symptomology in FXS were no longer related to one another in our sample.

These longitudinal analyses have specific implications for understanding the stability of autism symptoms in early childhood. Diagnosis of autism based upon two valid and reliable assessment measures appears to be stable in early childhood for young children with FXS. Further, while a relationship between lower cognition and increased autism symptoms may be evident in very young children with FXS, these variables may be no longer related in slightly older children, suggesting that a diagnosis is likely to remain stable and is not identified inaccurately in very young children, as a result of lower cognitive functioning. However, these results should be interpreted cautiously as they require replication in a larger sample of children.

[4] Some children had moved thus travel expense prohibited follow-up data collection. A few local families in this group reported difficulties with scheduling follow-up lab visits as a result of heavy school and therapy schedules.

IMPLICATIONS

Overall, these findings have specific implications for guiding intervention in the areas of language and social development for children with FXS. Other authors have suggested that children with different behavioral phenotypes benefit from different intervention approaches, based upon specific profiles of strengths and weaknesses (Dykens & Hodapp, 2001; Hodapp & Fidler, 1999). For example, young children with autism may need more interventions focused upon building receptive language skills, social development, and motivation to communicate, while children with FXS may derive the most benefit from interventions heavily targeting expressive language skills. Children with comorbid FXS and autism may need intense intervention in both receptive and expressive language areas with additional emphasis on developing social skills and communicative temptations for the child. Interventions, such as The Picture Exchange Communication System (PECS; Frost & Bondy, 1994), that are useful for motivating children with autism to communicate may provide useful for the same purposes in children with comorbid autism and FXS diagnoses.

Likewise, important social precursors for language development that are lacking in young children with autism may require treatment in children presenting with comorbid FXS and autism. Various early social skill intervention programs, such as The Denver Model (Rogers, 2005) or Relationship Development Inventory (RDI; Gutstein & Sheely, 2002), might be necessary in building up these foundational social skills necessary for subsequent language development. Further, while imitation might provide a viable tool for language and social learning in children with FXS, children with autism and co-morbid autism and FXS have phenotypic deficits in imitation abilities (Rogers et al., 2003). Thus, imitation skills make an important pre-requisite target skill for language intervention for children with autism as well as co-morbid autism and FXS. In addition, recognizing the relative strengths in receptive language in young children with FXS has many implications for educational and treatment strategies, such as providing a therapist and child with a common point of understanding in order to furnish rich and meaningful opportunities to learn and practice expressive activities.

These findings also have additional implications for assessment and evaluation practices in the FXS population. Inspection of the language profile of young children with known FXS diagnoses during diagnostic evaluations might help to signal the need for an additional screening for autism. While tests for FXS are objective and definitive, testing for autism is based on behavior and can be more difficult to ascertain, especially for professionals lacking extensive clinical experience with this population. The complexity of diagnosis is compounded when a child presents with FXS, a syndrome that presents with several features that may seem similar to those detected in autism. Given the finding that receptive language skills may distinguish children with autism from children with FXS, children with FXS who present with a severely depressed linguistic profile and limited receptive language growth over time should be further assessed for possible autism. Receptive language skills may provide important information for differential diagnosis of autism and fragile X. There are a number of tests on the market for the assessment of receptive and expressive language profiles in children under the age of three (See Table 3).

Table 3. Assessments for receptive and expressive language profiles in children under age 3.

Name of test	Age range assessed (in years:months)
The Mullens Scales of Early Learning (MSEL; Mullen, 1995)	0:0- 5:8
The Preschool Language Scale-4 (PLS-4; Zimmerman, Steiner, & Pond, 2002)	0:0- 6:11
Reynell Developmental Language Scales (Reynell & Gruber, 1978)	1:0- 6:9
Test of Early Language Development: Third Edition (TELD-3; Hresko, Reid, & Hammill, 1981)	2:0-7:11
Receptive-Expressive Emergent Language Test- Third Edition (REEL-3; Bzoch, League, & Brown, 2003)	0:0-3:0
Sequenced Inventory of Communication Development- Revised (SICD-R; Hedrick, Prather, & Tobin, 1984)	0:4- 4:0
Early Language Milestones Scale- Second Edition (ELM SCALE-2; Coplan, 1993)	0:0-3:0

Understanding the early language profiles of children presenting with autism, FXS, and co-morbid FXS and autism not only has implications for assessment and treatment practices in these populations, but further serves to clarify the behavioral phenotype of each of these populations. Of interest, while relatively strong receptive language skills in early development may make up the early communication phenotype for young boys with fragile X syndrome without autism, this may not be the case for young boys with FXS who also have autism (Roberts et al., 2001; Philofsky et al., 2004). Rather, children with co-morbid FXS and autism diagnoses look more like autism in that they do not show a relative strength in receptive language and include significant early social difficulties. While core social deficits are a vulnerability to children with an autism diagnosis, these capacities do not appear to be as significantly affected in young children with FXS alone. However, the receptive and expressive raw score difference in young children with co-morbid FXS and autism may look more like the FXS communication phenotype than the autism communication phenotype based on a significant discrepancy in expressive language skills acquisition compared to receptive language acquisition around the age of three (See Table 1 and Figure 1). Thus, elucidating the language and cognitive profiles of children with fragile X syndrome, autism, as well as co-morbid autism and fragile X syndrome provides an important area for both practitioners working with these populations, as well as researchers trying to understand the behavioral phenotypes associated with these disorders.

References

Abbeduto, L. & Hagerman, R.J. (1997). Language and Communication in Fragile X Syndrome. *Mental Retardation and Developmental Disabilities Research Reviews, 3:* 313-322.

Abbeduto, L., Murphy, M., Cawthon, S., Richmond, E., Weissman, M., Karadottir, S., & O'Brien, A. (2003). Receptive language skills of adolescents and young adults with Down and fragile X syndrome. *American Journal on Mental Retardation, 108(3): 149-160.*

American Psychiatric Association (2003). *Diagnostic and Statistical Manual of Mental Disorders-Fourth Edition, Text Revision.* Washington, DC: American Psychiatric Publishing.

Bailey, D., Hatton, D., & Skinner, M. (1998). Early developmental trajectories of males with fragile X syndrome. *American Journal on Mental Retardation, 103(1),* 29-39.

Bailey, D., Hatton, D., Skinner, M., & Mesibov, G. (2001). Autistic behavior, FMR1 protein, and developmental trajectories in young males with Fragile X syndrome. *Journal of Autism and Developmental Disorders, Vol 31(2),* 165-174.

Bailey, A., Phillips, W. & Rutter, M. (1996). Autism: towards an integration of clinical, genetic, neuropsychological, and neurobiological perspectives. *Journal of Child Psychiatry and Psychology, 37,* 89-126.

Bailey, D., Roberts, J., Hooper, S., Hatton, D., Mirrett, P., Roberts, J., & Schaaf, J. (2004). Research on Fragile X Syndrome and Autism: Implications for the Study of Genes, Environments, and Developmental Language Disorders. In M. Rice and S. Warren (Eds.), Developmental language disorders: From phenotypes to etiologies. (pp. 121-150). Mahwah, NJ: Lawrence Erlbaum Associates.

Baron-Cohen, S. (1989). The autistic child's theory of mind: A case of specific developmental delay. Journal of Child Psychology and Psychiatry, 30(2), *285-297.*

Bartak, L., Rutter, M., & Cox, A. (1975). A comparative study of infantile autism and specific developmental receptive language disorder: I. The children. British Journal of Psychiatry, 126, *127-145.*

Bartak, L., Rutter, M., & Cox, A. (1977). A comparative study of infantile autism and specific developmental receptive language disorders: III. discriminant function analysis. *Journal of Autism & Childhood Schizophrenia, 7(4),* 383-396.

Bennetto, L. & Pennington, B.F. (1996). The Neuropsychology of Fragile X Syndrome. In R.J. Hagerman & A. Cronister (Eds.), *Fragile X Syndrome: Diagnosis, Treatement, and Research, Second Edition.* Baltimore: The Johns Hopkins University Press.

Borghgraef, M., Fryns, J.P., Dielkens, A., Pyck, K., & Van den Berghe, H. (1987). Fragile (X) syndrome: A study of the psychological profile in 23 prepubertal patients. *Clinical Genetics, 32,* 179-186.

Burack, J., Hodapp, R., & Zigler, E. (1988). Issues in the classification of mental retardation: Differentiating among organic etiologies. *Journal of Child Psychology and Psychiatry,* 29, *765-779.*

Brown, W.T., Jenkins, E.C., Kuawczun, M.S., Wisniewski, K., Rudelli, R., Cohen, I.L., Fisch,

G., Wolf-Schein, E., Miezejeski, C., Dobkin, C. (1986). The fragile X syndrome. In H.M. Berg, K. Wisniewski, & D.A. Snider (Eds.), *Mental Retardevelopmentally delayedation: Research, Education and Technology Transfer* (Vol. 477, pp. 129-150). New York: New York Academy of Sciences.

Bzoch, K., League, R., & Brown, V. (2003). *Receptive-Expressive Emergent Language Test-Third Edition.* Austin, TX: PRO-ED.

Chakrabarti, S. & Fombonne, E. (2001). Pervasive developmental disorders in preschool children. *Journal of the Amaerican Medical Association*, 285(24), 3093-3099.

Coplan, J. (1993). *Early Language Milestone Scale- Second Edition.* Austin, TX- PRO-ED.

Cornish, K.M., Turk, J., Wilding, J., Sudhalter, V., Munir, F., Kooy, F., & Hagerman, R. (2004). Annotation: Deconstructing the attention deficit in fragile X syndrome: a developmental neurpsychological approach. *Journal of Child Psychology and Psychiatry, 45(6),* 1042-1053.

Dawson, G., Webb, S., Schellenberg, G., Dager, S., Friedman, S., Aylward, E., & Richards, T. (2002). Defining the broader phenotype of autism: Genetic, brain, and behavioral perspectives. *Development and psychopathology, 14,* 581-611.

Dykens, E.M. (1995). Measuring behavioral phenotypes: Provocations from the "new genetics." American Journal on Mental Retardation, 99, *522-532.*

Dykens, E. & Hodapp, R. (2001). Research in mental retardation: Toward an etiologic approach. Journal of Child Psychology and Psychiatry, 42(1), *49-71.*

Dykens, E., Hodapp, R., & Leckman, J. (1987). Strengths and weaknesses in the intellectual functioning of males with fragile X syndrome. American Journal of Mental Deficiency. Vol 92(2), 234-236.

Dykens, E., Hodapp, R. & Leckman, J. (1994). *Behavior and Development in Fragile X Syndrome.* Thousand Oaks, CA: Sage Publications, Inc.

Fein, D., Dunn, M., Allen, D.A., Aram, D.M., Hall, N., Morris, R. & Wilson, B.C. (1996). Language and Neuropsychological Findings. In I. Rapin (Ed.), *Preschool Children with Inadequate Communication.* London: MacKeith Press.

Fisch, G., Simensen, R., & Schroer, R. (2002). Longitudinal changes in cognitive and adaptive behavior scores in children and adolescents with the Fragile X mutation or autism. *Journal of Autism and Developmental Disorders, 32(2),* 107-114.

Fombonne, E. (1999). The epidemiology of autism: A review. *Psychological Medicine, 29,* 769-786.

Fombonne, E. (2005). The changing epidemiology of autism. *Journal of Applied Research in Intellectual Disabilities, 18(4),* 281-294.

Freund, L., Peebles, C., Aylward, E., & Reiss, A. (1995). Preliminary report on cognitive and adaptive behavior of preschool-aged males with fragile X. *Developmental Brain Dysfunction, 8,* 242-251.

Freund, L.S. & Reiss, A.L. (1991). Cognitive profiles associated with fra(X) syndrome in males and females. *American Journal of Medical Genetics, 38,* 542-547.

Frost, L. & Bondy, A. (1994). The Picture Exchange Communication System: Training Manual. Cherry Hill, NJ: Pyramid Educational Consultants, Inc.

Gutstein, S. & Sheely, R. (2002). Relationship Development Intervention with young children: Social and emotional development activities for Asperger Syndrome, autism, PPD and NLD. Philadelphia, PA: Jessica Kingsley Publishers, Ltd.

Hagerman, R. (1987). Fragile X syndrome. *Current Problems in Pediatrics, 17,* 627-674.

Hagerman, R. (1996). Fragile X syndrome. *Child and Adolescent Psychiatric Clinics of North America, Vol 5(4),* 895-911.

Hagerman, R. (1999). Clinical and molecular aspects of fragile X syndrome. In H. Tager-Flusberg (Ed), *Neurodevelopmental disorders* (pp. 27-42). Cambridge, MA: The MIT Press.

Hagerman, R., Ono, M., & Hagerman, P. (2005). Recent advances in fragile x: A model for autism and neurodegeneration. *Current Opinion in Psychiatry, 18,* 490-96.

Hanson, D.M., Jackson, A.W., & Hagerman, R.J. (1986). Speech disturbances (cluttering) in mildly impaired males with Martin-Bell/fragile X syndrome. *American Journal of Medical Genetics, 23(1-2),* 195-206.

Happe, F. & Frith, U. (1996). The neuropsychology of autism. *Brain: A Journal of Neurology. Vol 119(4),* 1377-1400.

Hedrick, D., Prather, E., & Tobin, A. (1984). *Sequenced Inventory of Communication Development (Revised Edition).* Seattle: University of Washington Press.

Hessl, D., Glaser, B., Dyer-Friedman, J., Blasey, C., Hastie, T., Gunnar, M., & Reiss, A. (2002). Cortisol and behavior in fragile X syndrome. *Psychoneuroendocrinology. Vol 27(7),* 855-872.

Hatton, D. D., Bailey, D.B., Hargett-Beck, M.Q., Skinner, M., Clark, R.D. (1999). Behavioral style of young boys with fragile X syndrome. *Developmental Medicine and Child Neurology, 41,* 625-632.

Hobson, R. (1993). Autism and the development of mind. Hillsdale, NJ, England: Lawrence Erlbaum Associates.

Hodapp R. & Fidler, D. (1999). Special education and genetics: Connections for the 21st Century. The Journal of Special Education, 33(3), *130-137.*

Hresko, W., Reid, D., & Hammill, D. (1981). The test of Early Language Development. *Autsin, TX: PRO-ED.*

Madison, L.S., George, C. & Moeschler, J.B. (1986). Cognitive Functioning in the Fragile X Syndrome: A Study of intellectual, memory and communication skills. *Journal of Mental Deficiency Research, 30*, 129-148.

Jarrold, C, Boucher, J., & Russell, J. (1997). Language profiles in children with autism. *Autism, 1*, 57-76.

Kau, A. S. M., Tierney, E., Bukelis, I., Stump, M., Kates, W., Trescher, W., & Kaufmann, W. (2004). Social behavior profile in young males with fragile X syndrome: characteristics and specificity. *American Journal of Medical Genetics, 126A:* 9-17.

Kaufmann, W. E., Cortell, R., Kau, A. S., Bukelis, I., Tierney, E., Gray, R., Cox, C., Capone, G., & Stanard, P. (2004). Autism spectrum disorder in fragile X syndrome: communication, social interaction, and specific behaviors. *American Journal of Medical Genetics 129A:* 225-234.

Kemper, M.B., Hagerman, R.J., & Altshul-Stark, D. (1988). Cognitive profiles of boys with fragile X syndrome. *American Journal of Medical Genetics, 30,* 191-200.

Kjelgaard, M. & Tager-Flusberg, H. (2001). An investigation of language impairment in autism: Implications for genetic subgroups. *Language and Cognitive Processes, 16(2/3),* 287-308.

Konstantareas, M.M. & Beitchman, J.H. Comorbidity of autistic disorder and specific developmental language disorder: Existing evidence and some promising future directions. In J.H. Beitchman, N.J. Cohen, M.M. Konstantereas, & R. Tannock. (Eds.), (1996). *Language, learning, and behavior disorders: Developmental, biological, and clinical perspectives* (pp. 178-196). New York, New York: Cambridge University Press.

Lee, A., Hobson, R. P., & Chiat, S. (1994). I, you, me, and autism: An experimental study. *Journal of Autism and Developmental Disorders, 24(2),* 155-176.

Levitas, A., Braden, M., Van Norman, K., Hagerman, R., McBogg, P. (1983). Treatment and intervention. In R.J. Hagerman & P.M. McBogg (Eds.), *The fragile X syndrome: Diagnosis, biochemistry, and intervention* (pp. 201-228). Dillon, CO: Spectra Publishing Co., Inc.

Lord, C., Risi, S., & Pickles, A. (2004). Trajectory of Language Development in Autistic Spectrum Disorders. In M. Rice and S. Warren (Eds.), *Developmental language disorders: From phenotypes to etiologies* (pp. 7-29). Mahwah, NJ: Lawrence Erlbaum Associates.

Marcus, L., Garfinkle, A., & Wolery, M. (2001). Issues in early diagnosis and intervention with young children with autism. In E. Schopler, N. Yurimiya, C. Shulman, & L. Marcus (Eds). *The Research Basis for Autism Intervention.* (pp. 135-150). New York: Kluwer Academic.

Massachusetts State Department of Education. (MDOE; 1998). Exploring options for young children with autism. *Malden, MA: Massachusetts Department of Education.*

McCauley, R.J. (2001). *Assessment of Language Disorders in Children.* Mahwah, New Jersey: Lawrence Erlbaum Associates, Inc.

Mervis, C. & Klein-Tasman, B. (2004). Methodological Issues in Group-Matching Designs: alpha Levels for Control Variable Comparisons and Measurement Characteristics of Control and Target Variables. *Journal of Autism and Developmental Disorders, 34(1),* 7-17.

Minshew, N., Meyer, J., Goldstein, G. (2002). Abstract reasoning in autism: A disassociation between concept formation and concept identification. *Neuropsychology, 16(3),* 327-334.

Mirret, P. L., Roberts, J. E., & Price, J. (2003). Early intervention practices and communication intervention strategies for young males with fragile X syndrome. Language Speech and Hearing Services in Schools, 34, 320-331.

Mullen, E. (1995). *Mullen Scales of Early Learning.* Circle Pines, MN: American Guidance Service, Inc.

Mundy, P., Sigman, M., Ungerer, J., Sherman, T. (1986). Defining the social deficits of autism: The contribution of non-verbal communication measures. *Journal of Child Psychology and Psychiatry, 27(5),* 657-669.

Newell, K., Sanborn, B. & Hagerman, R. (1983). Speech and Language Dysfunction in the Fragile X Syndrome. In R.J. Hagerman and P.M. McBogg (Eds.), *The Fragile X Syndrome: Diagnosis, Biochemistry, and Intervention.* Dillon, CO: Spectra Publishing Co., Inc.

Paul, R., Cohen, D,J., Breg, W.R., Watson, M. & Herman, S. (1984). Fragile X Syndrome: its relations to speech and language disorders. *Journal of Speech and Hearing Disorders, 49,* 328-332.

Paul, R., Dykens, E., Leckman, J.F., Watson, M., & Herman, S. (1987). A comparison of language characteristics of mentally retarded adults with fragile X syndrome and those with nonspecific mental retardation and autism. *Journal of Autism and Developmental Disorders, 17,* 457-468.

Pennington, B., O'Connor, R.A. & Sudhalter, V. (1991). Toward a Neuropsychology of Fragile X Syndrome. In R.J. Hagerman & A.C. Silverman (Eds.), *Fragile X Syndrome: Diagnosis, Treatment, and Research.* Baltimore: The Johns Hopkins University Press.

Philofsky, A., Hepburn, S. L., Hayes, A., Hagerman, R., & Rogers, S. J. (2004). Linguistic and cognitive functioning and autism symptoms in young children with fragile X syndrome. *American Journal on Mental Retardation, 109(3),* 208-218.

Rapin, I. (1996). *Preschool Children with Inadequate Communication.* London: MacKeith Press.

Reiss, A.L. & Dant, C.D. (2003). The behavioral neurogenetics of fragile x syndrome: Analyzing gene-brain-behavior relationships in child developmental psychopathologies. *Development and Psychopathology, 15,* 927-968.

Reynell, J. & Gruber, C. (1978). *Reynell Developmental Language Scales- US edition.* Los Angeles, CA: Western Psychological Services.

Rice, M., Warren, S., & Betz, S. (2005). Language symptoms of developmental language disorders: An overview of autism, Down syndrome, fragile X, specific language impairment, and Williams syndrome. *Applied Psycholinguistics, 26(1),* 7-27.

Roberts, J.E., Mirrett, P. & Burchinal, M. (2001). Receptive and Expressive Communication Development of Young Males with Fragile X Syndrome. *American Journal on Mental Retardation, 106,* 216-30.

Rogers, S. J. (2005). Play Interventions for Young Children with Autism Spectrum Disorders. In L. Reddy, T. Files-Hall, & C. Schaefer (Eds.) Empirically based play interventions for children. (pp. 215-239). Washington, DC: American Psychological Association.

Rogers, S., Hepburn, S., Stackhouse, T., & Wehner, E. (2003). Imitation performance in toddlers with autism and those with other developmental disorders. *Journal of Child Psychology and Psychiatry, 44(5),* 763-781.

Rogers, S.J., Wehner, E.A., & Hagerman, R. (2001). The behavioral phenotype in fragile X: Symptoms of autism in very young children with fragile X syndrome, idiopathic autism, and other developmental disorders. *Journal of Developmental and Behavioral Pediatrics,22 (6),* 409-417.

Sherman, S. (2002). Epidemiology. In R. J. Hagerman & P.J. Hagerman (Eds.), *Fragile X syndrome: Diagnosis, treatment, & research* (3rd ed.). Baltimore, MD: The Johns Hopkins University Press.

Siegel, D., Minshew, N., & Goldstein, G. (1996). Wechsler IQ profiles in diagnosis of high-functioning autism. *Journal of Autism and Developmental Disorders, 26(4),* 389-406.

Sigman, M. (1994). What are the core deficits in autism? In S. Broman & J. Grafman (Eds.), Atypical cognitive deficits in developmental disorders: Implications for brain function. (pp. 139-157). Hillsdale, NJ: Lawrence Erlbaum.

Sudhalter, V., Cohen, I.L., Silverman, W., Wolf-Schein, E. (1990). Conversational analyses of males with fragile X syndrome, Down Syndrome, and autism: Comparison of the emergence of deviant language. *American Journal of Mental Retardation, 94,* 431-441.

Szatmari, P. (2000). The classification of autism, Asperger's syndrome, and pervasive developmental disorder. *Canadian Journal of Psychiatry, 45(8),* 31-738.

Tager-Flusberg, H. (2000). The Challenge of Studying Language Development in Children With Autism. In L. Menn and N. Bernstein-Ratner(Eds.), *Methods for Studying Language Production.* Malwah, NJ: Lawrence Erlbaum and Associates.

Turk. J. & Graham, P. (1997). Fragile X syndrome, autism, and autistic features. *Autism, 1(2),* 175-197.

Turner, G., Daniel, A. & Frost, D.A. (1980). X-linked Mental Retardation, Macro-orchidism and the (X)(q27) Fragile Site. *Journal of Pediatrics, 96,* 837-841.

Volden, J. & Lord, C. Neologisms and idiosyncratic language in autistic speakers. *Journal of Autism and Developmental Disorders, 21(2),* 109-130.

Volkmar, F., Klin, A., Siegel, B., Szatmari, P., Lord, C., Campbell, M., Freeman, B.J., Cicchetti, D.V., Rutter, M., & Kline, W. (1994). Field trial for autistic disorder in DSM-IV. *American Journal of Psychiatry, 151(9),* 1361-1367.

Wolf-Schein, E.G., Sudhalter, V., Cohen, I.L., Fisch, G.S., Hanson, D., Pfadt, A.G., Hagerman, R., Jenkins, E.C. & Brown, W.T. (1987). Speech-Language and the Fragile X Syndrome: Initial Findings. *ASHA, 29,* 35-38.

Yeargin-Allsopp, M., Rice, C., Karapurkar, T., Doernberg, N., & Murphy, C. (2003). Prevalence of autism in a US metropolitan area. *Journal of the American Medical Association, 289(1),* 49-55.

Zigler, E. (1969). Developmental versus difference theories of mental retardation and the problem of motivation. *American Journal of Mental Deficiency, 73(4),* 536-556.

Zimmerman, I., Steiner, V., Pond, R. (2002). *Preschool Language Scale- Fourth Edition.* San Antonio, TX: The Psychological Corporation.

In: New Research on Fragile X Syndrome
Editor: Jeffrey J. Upner, pp. 23-35

ISBN: 1-60021-281-6

Chapter 2

DOES COGNITIVE PERFORMANCE INCREASE OR DECREASE ACROSS DEVELOPMENT IN FRAGILE X SYNDROME: IMPACT OF VERBAL MENTAL AGE AND GENDER

Kim Cornish[5]
Depts. of Educational Psychology & Neurology/Neuropsychology Montreal, Canada
John Wilding
Royal Holloway University of London, Egham Hill, Egham, Surrey TW20 0EX, UK

ABSTRACT

The present study assessed cognitive functioning in sixty participants with fragile X syndrome (30 males and 30 females), ranging in age from 6 years to 34 years, divided into two separate groups distinguished by Chronological Age (under 16 years and over 16 years). Intellectual level ranged from 3 years to 13 years in affected males and from 3 years to 20 years in affected females. Performance was compared to typically developing participants matched on intellectual level and ranging in chronological age from 4 years to 14 years. A comprehensive battery of cognitive measures was employed with both groups, covering the domains of language, articulation, and spatial cognition (visual-perceptual, visuo-spatial and visuo-construction).

Two important findings emerged: First, contrary to previous research, neither fragile X males nor females displayed a reduction in performance with increasing age across any of the cognitive measures. Indeed, performance on all tasks improved with verbal mental age (VMA) to the same degree in both fragile X and typically developing groups. Second, we found evidence of a gender-specific fragile X 'signature' whereby males and females of similar verbal mental age display different cognitive profiles, with males especially vulnerable to impairments on skills that required a visuo-motor component. Together, these findings

---

[5] Address for correspondence: K.M. Cornish, McGill University, 3700 McTavish Street, Montréal, H3A 1Y2, Canada , Email: kim.cornish@mcgill.ca

represent a unique contribution to the existing literature. The theoretical and practical implications will be discussed.

INTRODUCTION

This past decade has seen unparalleled advances in our understanding of how genes can impact on development across a number of different levels from the genetic and brain levels to overt behaviour with complex effects all the way from infancy through to adulthood. Collaboratively, this new knowledge gathered from the disciplines of molecular genetics, neuro-imaging, and developmental neuroscience has begun a careful delineation of the normal trajectory of function across the lifespan. These advances have also paved the way for more detailed investigations of atypical trajectories, providing developmentalists with a unique 'window' on the cognitive phenotypic outcomes of disorders with differing genetic etiologies. Indeed, numerous studies now attest to the distinct cognitive end-states or 'signatures' that characterise given genetic disorders such as Williams syndrome[1], Prader-Willi syndrome[2], Turner's syndrome[3], Down Syndrome[4] and others. Although these phenotypic end-states yield important information that help target syndrome-specific clinical and educational programs, few studies have incorporated a developmental approach towards understanding atypical pathways, and fewer still have looked at specific adult phenotypes to see if changes in outcomes have remained stable or changed across developmental time.

The focus of the present study is to examine the pattern of cognitive functioning in childhood and adulthood in a disorder that has received considerable recognition in the last 10 years – fragile X syndrome (FXS) [5,6,7,8]. FXS is a fascinating syndrome, whose single gene aetiology affords us the unique opportunity to begin to understand the relationships among genes, brain and behaviour. The condition is the most prevalent known inherited cause of developmental delay in humans, affecting 1 in 4000 males and 1 in 8000 females.[9,10,11] Physical features include elongated face, large prominent ears and forehead, and, in males, macro-orchidism (post-puberty).[12] However, the wide variability in expression, in both males and females, makes diagnosis on physical features alone almost impossible. The most defining feature in FXS, especially in boys with the condition, is mental retardation and the resulting behavioral phenotype. It is precisely because of their quite normal appearance that many affected children are not recognised until relatively later in development as having FXS, which is why a close examination of the phenotypic signature and its developmental time-line is crucial in helping clinicians towards an early diagnosis of FXS.

As stated above, by virtue of its single gene aetiology, FXS represents an important model for understanding the impact of gene expression on the development and normal functioning of the central nervous system. While it is becoming commonplace to investigate the phenotypic effects resulting from the loss of a single gene product in animal models, few naturally occurring genetic anomalies exist in humans that allow us to examine the specific contribution of a single gene to behaviour. FXS is caused by an expansion of the CGG repeat at the beginning of the FMR-1 (fragile X mental retardation 1) gene on the X chromosome. In most clinically affected individuals, there is a significant expansion of repeats (>200), resulting in the silencing of the FMR1 gene, and a lack or absence of the protein product carried by the FMR1 gene. It is the absence of this fragile X protein during early brain development that results in the characteristic phenotype associated with this syndrome. Due to X-linkage, almost all affected males present with mental retardation compared to approximately a third of affected females.

At the brain level, there have been a number of interesting and illuminating findings but the majority of research has focused exclusively on the adult brain. Studies have revealed decreased size of the posterior vermis of the cerebellum in males and females.[13,14] Other structures affected by *FMR1* status include the caudate nucleus [15] and the hippocampus.[16,17] In addition, several studies have reported a correlation between neuroanatomical abnormalities and the degree of functional impairment in the full mutation. For example, posterior vermis volumes are positively correlated with performance on specific measures of intelligence, visuo-spatial ability, and so-called 'Executive Function' (an umbrella term often used to describe a series of activities that include planning, working memory and set shifting), suggesting a putative functional role for this structure.[13] Taken together the purely structural and combined structural/functional studies implicate the cerebellum, caudate nucleus, and hippocampus as potential sites for phenotypic effects from abnormal *FMR1* gene expression. However, it is important to stress that abnormal structures form part of widely distributed networks and that deficits may be observed across a broad range of activities. Future research will undoubtedly locate more fine-tuned interconnections at the cellular level that impact on early brain development and the subsequent development of cognitive functions.

In the following sections we will highlight some of the main advances in our understanding of the cognitive phenotype of FXS in both males and females, and then proceed to address our aim of understanding how the childhood phenotype develops into adulthood. At the intellectual level, the majority of FXS males (> 95%) will display IQ's within the moderate-severe range of impairment, with profiles emerging as young as three years of age.[18] In females, the phenotypic variation is such that some girls only show sub-clinical learning disabilities [19] whilst approximately 50% display more moderate - severe mental retardation similar in profile to boys with FXS. The X-inactivation status (ratio of normal alleles on the X chromosome) in FXS females is seen as the major contributor to the heterogeneity of intellectual disability and neuropsychological deficits. Surprisingly, however, few studies have addressed the issue of whether degree of intellectual impairment changes with age, albeit progressing at a slower pace compared to typical developing children, or whether intellectual level actually decreases with increasing age, or even remains static across developmental time. Moreover, does change in intelligence functioning correlate with performance across cognitive domains? Although published studies have consistently observed a decline in cognitive abilities from middle-late childhood onwards,[20,21,22] Cornish, Turk, Wilding et al (2004) have previously argued that this decline might possibly be due to the fragile X's child's increasing problems in maintaining and developing successful cognitive strategies that keep pace with their age-normed cohort rather than an actual regression in intellectual level or failure in neural development. However, few if any studies have examined intellectual decline and its relation to cognition in females and fewer still have examined any profile beyond late childhood.

In the present study, we sought to examine cognitive functioning and the trajectory of performance across a range of tasks that tap known weaknesses and strengths in the fragile X profile. An important point to recognise is that fragile X, *at least by mid childhood and in males*, is not characterized by global mental retardation. Instead, the syndrome 'signature' is one of uneven abilities within and across cognitive domains. For example, difficulties in speech fluency characterised by repetitive and impulsive speech [23] are often accompanied by difficulties in grammar and pragmatics [19] with many fragile X children only reaching the level of MA matched control children. However, receptive language and comprehension skills are not as affected. In the spatial domain, we see a similar pattern of weakness for skills that require abstract construction but relatively good skills for tasks that require more meaningful identification and construction. Other cognitive domains such as working memory and attention are also severely affected in fragile X syndrome.[24,25,26,27] In contrast to the well-documented profile of fragile X males, few studies have addressed working memory in affected girls and those that have focused almost exclusively on adult women. In the few studies to date on females, there is emerging evidence for a more subtle impairment in working memory thresholds, [28,29] visual spatial skills especially for abstract information, [30] and difficulties in initiating conversational interactions.[31]

Here we attempt to move our understanding of the fragile X cognitive profile in two new directions. First by examining sex differences in performance on the same tasks across three cognitive domains: spatial cognition, language, and articulation; and second, by looking at how developmental maturity, using a wider age range than used in previous studies, impacts on cognitive functions across domains and across gender.

METHOD

Participants

A total of 60 individuals with fragile X syndrome constituted the final sample for analyses (30 males and 30 females). Age ranged from 6 years to 34 years, divided into two separate groups distinguished by Chronological Age (CA under 16 years and over 16 years). Intellectual level, as measured by the British Picture Vocabulary Scales (BPVS) [32] which produces a verbal mental age, ranged from 3 years to 13 years in affected males and from 3 years to 20 years in affected females. Performance was compared to 60 typically developing participants matched on Verbal Mental Age (VMA) and therefore chronologically younger than the fragile X participants. Age ranged from 4 years to 14 years.

All fragile X participants were recruited from the UK fragile X Society and all have previously received a DNA analysis confirming the FMR1 full mutation (>200 CGG repeats).

Design and Measures

Six age-appropriate tasks that tapped the domains of spatial cognition (visuo-spatial, visuo-motor, and visual perceptual), language and articulation were completed by all participants.

Spatial Cognition:

The *Block Design task* [33] is a visuo-spatial task in which cubes with red, white or mixed faces are arranged to construct target abstract two-dimensional patterns. Scores in each case are dependent upon the elapsed time for the correct construction of the designs.

The *Triangles task* [34] is similar to the Block Design task but instead of using red\white cubes to construct target, abstract designs, triangles (blue on one side, yellow on the other) are used. As on the Block Design task, scores in each case are dependeny upon the elapsed time for the correct construction of the designs.The *Draw-A-Person task* [35] is a visuo-constructional task that requires a graphic response to stored structure information, in this case an animate object. A time limit of one minute is given to complete the drawing. The *Gestalt Closure task* [34] is a visuo-perceptual task that requires the gestalt integration of patterns of black against white in order to recognise common objects.*Language*

The *Test of the Reception of Grammar* (TROG) [36] which assess the comprehension of grammar using a picture pointing response format that eliminates the need for expressive language ability. Each participant is presented with an array of four pictures and has to indicate which one represents the sentence spoken by the examiner, for example, "he is sitting on the table" as opposed to the grammatical distracter "she is sitting on the table". This test is concerned with understanding parts of speech, simple and complex sentences, pronouns, word inflections, relative clauses, and embedding. The TROG also allows the calculation of an age equivalent score.

The *Goldman-Fristoe Test of Articulation* (GFTA) [37] is a test of articulation and measures spontaneous production of all except one of the English consonant sounds as well as 11 consonant blends.

Procedure

All children were tested individually in a quiet room in a familiar environment whether at school or at home, in two 35-minute sessions with a 15-minute interval.

RESULTS

Intellectual Level in Fragile X Males and Females

Preliminary Analyses of Variance indicated a marked discrepancy in verbal mental age (VMA) between older male and female FXS participants ($F(1,27) = 18.45$, $p < .001$), with adult female FXS participants demonstrating higher VMA than FXS males (despite selecting chronologically older FXS males in an attempt to achieve a match). These differences were therefore also present in the matched control groups (for the whole older sample sex difference $F(1,55) = 32.54$). It was therefore decided to drop female participants with VMA scores higher than the highest scoring male (VMA = 160 months) and to divide the remaining sample at the median VMA (81 months), thus creating two groups matched overall on mean VMA. Necessarily within these groups FXS and typically developing controls were not matched on CA, since the latter were younger than their FXS counterparts. Table 1 gives the mean VMA and CA for all subgroups, together with the number of participants, and demonstrates that there was close matching of VMA between FXS and Controls and between Males and Females within each level of VMA. CA was also similar for males and females within the FXS groups and within the control group with lower VMA, but there were wide variations in mean CA between the different subgroups at the higher VMA level. This issue will be addressed in the statistical analyses. The total number of participants in the selected groups was 51 in the FXS group and 55 in the Control group.

Scores from all six tasks were converted to standard deviates to ensure that results from all tasks could be presented on the same scale. Multivariate Analysis of Covariance was carried out with three Between Subject factors (Group with two levels: FXS and Control; Gender with two levels: Male and Female; VMA Level with two levels: below 81 and above 80). CA was added as a covariate, for the reasons already cited above.

VMA and CA

The analysis of covariance showed that both CA and VMA were related significantly to overall performance ($F(6,82) = 7.02$ and 8.92, $p < .001$) but, while VMA was significantly related to each task separately ($F(1,82)$ ranged from 6.31 to 40.31, $p < .01$ in all cases), CA was only related to GFTA performance ($F(1, 87) = 33.4$, $p < .001$).

There were no significant interactions of VMA with either of the other variables (Group and Gender), indicating that improved performance with increasing VMA did not differ in the four subgroups in any of the tasks. Table 2 gives the mean of the six standardised scores for the four subgroups at each level of VMA. Though there is an indication that the female FXS group improved less than the other three groups, the relevant interactions again did not approach significance.

Table 1. Mean VMA and CA values in each group.

	FXS		Control	
	Males	Females	Males	Females
VMA < 81				
VMA	66	62	62	61
CA	137	133	70	64
N	17	13	9	7
VMA > 80				
VMA	122	119	107	115
CA	287	168	94	128
N	12	9	20	19

Table 2. Mean standardised scores and standard deviations over all tasks for each group at each level of VMA.

	VMA < 81	VMA > 80
FXS Male	-1.03 (.47)	.10 (.44)
FXS Female	-.60 (.66)	.21 (.71)
Control Male	-.10 (.58)	.69 (.30)
Control Female	-.41 (.65)	.75 (.32)

FXS v. Control and Male v. Female Differences

The main focus of interest was differences between the FXS and Control group and between the Gender groups. In order to carry out Newman-Keuls comparisons for each task in a further Multivariate Analysis of Variance each score was regressed on CA to remove differences due to this variable and the residual score calculated. Multivariate Analysis of Variance was then carried out on these residual scores and Newman-Keuls comparisons carried out to detect which of the four subgroups differed significantly from each other (the criterion was set at the 5% level). The significant group differences are shown in Table 3 and Figure 1 plots the means for each subgroup over each of the six tasks.

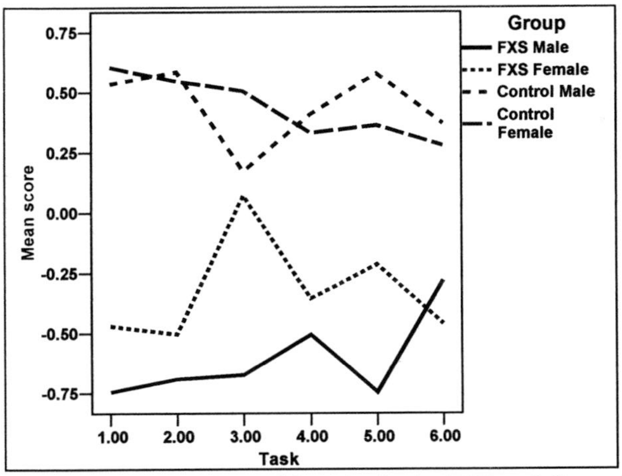

Figure 1. Mean scores on each task for males and females in the FXS and Control Groups. Task 1 = Triangles K-ABC; Task 2 = WPPSI Block Design; Task 3 = Draw a Person; Task 4 = TROG; Task 5 = GFTA; Task 6 = Gestalt

Table 3: Pattern of differences across groups.

Triangles	FXMales < FXFemales < ControlMales = ControlFemales
Block Design	FXMales = FXFemales < ControlMales = ControlFemales
DAP	FXMales < FXFemales = ControlMales = ControlFemales
TROG	FXMales = FXFemales < ControlMales = ControlFemales
GFTA	FXMales < FXFemales < ControlMales = ControlFemales
Gestalt	FXMales = FXFemales < ControlMales = ControlFemales

To summarise the significant findings:

1. Performance on all tasks improved with VMA to the same degree in all the subgroups.
2. Male FXS groups were worse than the corresponding Control groups on all tasks.
3. Female FXS were worse than the corresponding Control groups on all tasks except DAP.
4. Male FXS were worse than Female FXS on Triangles, DAP and GFTA.

Discussion

The main aim of the present study was to assess the range of impairment in children and adults with fragile X syndrome across a battery of cognitive tasks that tap spatial cognition, language and articulation and to assess the extent of any gender differences in performance and to what extent intellectual level impacts upon performance. Although there have been numerous studies in the past few years that have attempted to define the phenotypic outcomes associated with fragile X syndrome, [23,28,38,39] most have concentrated solely on childhood outcomes. Relatively few studies have looked exclusively at adulthood, and of these fewer still have focused on comparisons of both genders [40] Our study represents an important addition to the literature by providing novel data that compare cognitive performance in affected males and females across the lifespan, in childhood and adulthood, and using the same tasks. Moreover, we examine performance across gender and age group within the context of changes in intellectual level. Two important findings emerged:

1. Neither fragile X males nor females displayed a reduction in performance with increasing age across any of the cognitive measures. Indeed, all groups improved as VMA increased although such increases will also have been strongly associated with CA increase. These increases were not significantly different in the typically developing groups and the FXS groups (there were no significant interactions of Group and VMA level). These findings are important at a number of levels. At the *cognitive level*, they indicate that individuals with fragile X do not reach a 'learning threshold' in late childhood/early adolescence but continue to improve cognitively with increasing verbal mental age, be they children or adults. This finding contrasts with previous studies that have suggested a decline in function, notably in males from late childhood or adolescence onwards.[20,21,22] However, at first glance, our findings, although seemingly contradictory to those cited above, are quite different in nature. For example, Fisch and colleagues used traditional, standardised measures of IQ (e.g. Stanford- Binet) to assess decline across childhood into adolescence. In contrast, our present study used more task specific measures to assess differing aspects of cognitive ability rather than assessing a global skill or domain. In addition, unlike previous research that has focused on observing age-related changes specifically within the child-adolescence period, our study covered performance into adulthood. Crucially, however, we examined age-related changes in cognitive functioning as a function of increasing verbal mental age, rather than using age-

normed IQ composite scores. Our findings indicate when individuals with fragile X are matched on VMA to typically developing individuals then performance will improve at the same rate with increasing maturity, though always performing at a lower level. In other words, absolute level of performance is affected by the genetic disorder, but the development is at a normal rate. This has important implications for educational and clinical interventions, not least because this new information suggests that a plateau is not reached by early adulthood. Instead cognitive development continues to increase beyond adolescence no more slowly than normally developing groups matched on VMA. Future research using a prospective longitudinal design to chart developmental pathways from childhood into early adulthood will be needed to further define the rate of age-related changes in fragile X males and females both across and within cognitive domains.

2. Our findings indicate a possible gender-specific Fragile X 'signature' whereby males and females of the similar verbal mental age display different cognitive profiles. Fragile X males as a group performed worse overall, followed by fragile X females and then typically developing controls. When gender differences emerged they were only within the fragile X groups. Females were superior to males on tasks that tapped visuo-spatial skill (Triangles), visuo-constructive skill (Draw A Person), and articulation (Goldman-Fristoe Test of Articulation) but both genders displayed comparable performance on tasks that involved gestalt integration (Gestalt Closure) and language comprehension (Test of the Reception of Grammar). The pattern of *gender differences* in fragile X represents a unique and important finding that highlights a specific male weakness for skills that depend on visual and spatial construction of abstract, meaningless designs to form a whole, and for recall and construction of a familiar design. The fact that both tasks require relatively good visuo-motor integration for successful completion may be the key to understanding the fragile-X male deficit. In affected boys, there have been several reports of sensori-motor integration difficulties and visual-motor impairments, [25,41] with specific difficulties in performance for tasks requiring the integration of visual information for effective motor control.[30] In contrast, relative strengths have been reported on visuo-perceptual tasks, such as gestalt integration, and face recognition tasks.[42,43] One possible explanation for this dissociation has come from a series of recent neurobiological studies by Kogan and colleagues that speculate that FXS is associated with selective deficits in magnocellular/dorsal stream visual functioning which result in an underlying impairment in regions of the brain devoted to processing visual information for the purpose of control of action. In contrast, the parvocellular/ventral stream is a transformer of visual input necessary for the conscious perception of visual information (e.g. color and recognition).[44,45] Because fragile X is caused by a single gene, Kogan et al proposed that the protein normally expressed by the FMR1 gene and lacking in FXS protein (FMRP) might be more important to normal magnocellular neuron function or structure than parvocellular function. Indeed, post-mortem analyses did reveal FMRP to be expressed in greater abundance in the magnocellular (M) pathway of monkey and human brains suggesting that this pathway is more reliant on the protein for normal functioning. In contrast, post-mortem analyses of a fragile X adult patient revealed abnormal neuromorphology of LGN neurons such that they appear more like

parvocellular (P) neurons than M-neurons. Kogan et al concluded that abnormal visual-motor behavior in fragile X can be partially attributed to an M-pathway deficiency in FMRP. In affected females, due to X inactivation, there may be some levels of FMR protein within the magnocellular pathway that is absent in males. However, whilst these levels may be sufficient to produce a female superiority in terms of visuo-motor performance, the levels are clearly insufficient to ensure comparable performance to that of non-affected, typically developing individuals. These findings highlight the need for clinical treatments to focus on improving early visual-motor co-ordination. Further research is now needed to determine whether a similar neurobiological substrate is impacting on articulation abilities which were also differentially impaired between fragile X males and females.

In conclusion, a number of novel and important findings have emerged from this study. First, age-related increases in cognitive functioning are clearly evident, and although fragile X performance never matches that of Control groups matched on VMA it does improve at the same rate from a lower baseline. Second, the emergence of a gender specific fragile X 'signature' that is likely to reflect the differing levels of FMR protein in brain regions that are especially vulnerable in affected males and females, for example a selective disruption in the magnocellular visual pathway leading to specific visuo-motor deficits. Together, these findings demonstrate the importance of finer-tuned analyses of cognitive functioning (not solely IQ performance) across the life-span that will tease apart subtle differences in developmental trajectories that are due to fragile X syndrome and those that result from the generalised effects of intellectual impairment. With greater knowledge of the *fragile X signature* comes the need to target interventions, both educational and clinical, that recognise the developmental nature of the cognitive profiles of affected males and females. It is therefore crucial that interventions begin as early in development as possible but continue across the lifespan into adulthood in order for individuals with fragile X and their families to benefit from these tremendous advances in our knowledge of a condition that affects millions world-wide.

REFERENCES

[1] Donnai, D., Karmiloff-Smith, A. (2000). Williams syndrome: from genotype through to the cognitive phenotype. *American Journal of Medical Genetics, 97,* 164-171.

[2] Whittington, J., Holland, A., Webb, T., Butler, J., Clarke, D., & Boer, H. (2004) Academic underachievement by people with Prader-Willi syndrome. *Journal of Intellectual Disability Research, 48,* 188-200.

[3] Nijhuis-Van der Sanden, M.W., Eling, P.A., & Otten, B.J. (2003). A review of neuropsychological and motor studies in Turner Syndrome. *Neuroscience and Bio-behavioral Review,27,* 329-338.

[4] Wilding, J., & Cornish, K.M. (2004). Attention and executive functions in children with Downs Syndrome. In J. Mallard (Ed.), *Focus on Down Syndrome Research.* Nova Publishing.

[5] Cornish, K. M., Levitas, A., & Sudhalter, V. (in press). Fragile X syndrome: the journey from genes to behaviour. In M. M. Mazzocco (Ed.), *Neurogenetic Developmental Disorders: Manifestation and Identification in Childhood* (pp.1-50). New York: MIT Press.

[6] Cornish, K. M., Turk, J., Wilding, J., Sudhalter, V., Kooy, F., Munir, F., et al. (2004). Annotation: Deconstructing the attention deficit in Fragile X syndrome. *Journal of Child Psychology and Psychiatry, 45,* 1042-1053.

[7] Hessl, D., Dyer-Friedman, J., Glaser, B., Wisbeck, J., Barajas, R.G., Taylor, A., & Reiss, A.L. (2001). The influence of environmental and genetic factors on behavior problems and autistic symptoms in boys and girls with fragile X syndrome. *Pediatrics, 108,* E88.

[8] Turk J. (1998). Fragile X syndrome and attentional deficits. *Journal of Applied Research in Intellectual Disabilities, 11*, 175-191.

[9] De Vries, B.B., Halley, D.J., Oostra, B.A., & Niermeijer, M.F. (1998). Fragile X syndrome. *Journal of Medical Genetics, 35,* 579-589.

[10] Kooy, R.F., Willesden, R. & Oostra, B.A. (2000). Fragile X Syndrome at the turn of the century. *Molecular Medicine Today, 6*, 193-198.

[11] Turner, G., Webb, T., Wake, S., & Robinson, H. (1996) Prevalence of fragile X syndrome. *American Journal of Medical Genetics, 64*, 196-197.

[12] Lachiewicz, A.M., Dawson, D.V., & Spiridigliozzi, G.A. (2000). Physical characteristics of young boys with fragile X syndrome: reasons for difficulties in making a diagnosis in young males. *American Journal of Medical Genetics, 92,* 229-236.

[13] Mostofsky, S.H., Mazzocco, M.M., Aakalu, G., Warsofsky, I.S., Denckla, M.B., & Reiss, A.L. (1998). Decreased cerebellar posterior vermis size in fragile X syndrome: Correlation with neurocognitive performance. *Neurology, 50*, 121-130.

[14] Reiss, A., Alyward, E., Freund, L., Joshi, P.K., & Bryan, R.N. (1991). Neuroanatomy of Fragile X Syndrome: the posterior Fossa. *Annals of Neurology, 29*, 26-32.

[15] Eliez, S., Blasey, CM., Freund, L.S., Hastie, T., & Reiss, A.L. (2001). Brain anatomy, gender and IQ in children and adolescents with fragile X syndrome. *Brain, 24*, 1610-1618.

[16] Kates, W.R., Abrams, M.T., Kaufmann, W.E., Breiter, S.N., & Reiss, A.L. (1997). Reliability and validity of MRI measurement of the amygdala and hippocampus in children with fragile X syndrome. *Psychiatry Research, 75*, 31-48.

[17] Reiss, A.L., Lee, J., & Freund, L. (1994). Neuroanatomy of fragile X syndrome: The temporal lobe. *Neurology, 44*, 1317-1324.

[18] Skinner, M., Hooper, S., Hatton, D.D., Roberts, J., Mirrett, P., Schaaf, J., Sullivan, K., Wheeler, A., & Bailey, D.B. Jr. (2005) Mapping nonverbal IQ in young boys with fragile X syndrome. *American Journal of Medical Genetics, Part A, 132,* 25-32.

[19] Bennetto, L., & Pennington, B.F. (2002). Neuropsychology. In R. J. Hagerman & P. J. Hagerman (Eds.), *Fragile X Syndrome: Diagnosis, Treatment, and Research, 3rd Edition* (pp. 206-248). Baltimore, MD: Johns Hopkins University Press.

[20] Fisch, G.S., Simensen, R.J., & Schroer, R.J. (2002) Longitudinal changes in cognitive and adaptive behavior scores in children and adolescents with the fragile X mutation or autism. *Journal of Autism and Developmental Disorders, 32* , 107-114.

[21] Fisch, G.S., Carpenter, N.J., Holden, J.J., Simensen, R., Howard-Peebles, P.N., Maddalena, A., Pandya, A., & Nance, W. (1999). Longitudinal assessment of adaptive and maladaptive behaviors in fragile X males: growth, development, and profiles. *American Journal of Medical Genetics, 83,* 257-263.

[22] Fisch, G.S., Shapiro, L.R., Simensen, R., Schwartz, C.E., Fryns, J.P., Borghgraef, M., Curfs, L.M., Howard-Peebles, P.N., Arinami, T., & Mavrou, A. (1992). Longitudinal changes in IQ among fragile X males: clinical evidence of more than one mutation? *American Journal of Medical Genetics, 43,* 28-34.

[23] Belser, R.C, & Sudhalter, V. (2001) Conversational Characteristics of Children with Fragile X Syndrome: Repetitive Speech. *American Journal of Mental Retardation, 106,* 28-38.

[24] Cornish, K. M., Sudhalter, V., & Turk. J. (2004). Attention and speech in Fragile X syndrome. *Mental Retardation and Developmental Disabilities Research Reviews, 10,* 11-16.

[25] Freund, L.S., & Reiss, A.L. (1991). Cognitive profiles associated with the fra(X) syndrome in males and females. *American Journal of Medical Genetics, 38,* 542-547.

[26] Munir, F., Cornish, K.M., & Wilding, J. (2000). Nature of working memory deficit in Fragile X syndrome. *Brain and Cognition, 44,* 387-401.

[27] Wilding, J., Cornish, K. M., & Munir, F. (2002). Further delineation of the executive deficit in males with Fragile X syndrome. *Neuropsychologia, 40,* 1343-1349.

[28] Mazzocco, M.M., Bhatia, N.S., & Lesniak-Karpiak, K. (2005). Visuo-spatial deficits in fragile X or Turners Syndrome. *Child Neuropsychology.* In press.

[29] Kirk, J.W., Mazzocco, M.M., & Kover, S.T. (2005). Assessing executive dysfunction in girls with fragile X or Turner syndrome using the Contingency Naming Test (CNT). Developmental Neuropsychology, 28, 755-777.

[30] Cornish, K.M., Munir, F., & Cross, G. (1999). Spatial cognition in males with Fragile X Syndrome: evidence for a neuropsychological phenotype. *Cortex, 35,* 263-271.

[31] Lesniak-Karpiak, K., Mazzocco, M.M., & Ross, J.L. (2003). Behavioral assessment of social anxiety in females with Turner or fragile X syndrome. *Journal of Autism and Developmental Disorders, 33,* 55-67.

[32] Dunn, L., Dunn, P., Whetton, C. & Pintillie, D. (1982) British Picture Vocabulary Scale. Windsor, NFER.

[33] Wechsler, D. (1992). Wechsler Intelligence Scale for Children, 3rd edn, UK. The Psychological Corporation, Harcourt Brace and Company.

[34] Kaufman, A. S. & Kaufman, N. L. (1983). Kaufman Assessment Battery for Children. Circle Pines, MN: American Guidance Service.

[35] Naglieri, J. A., *Draw-A-Person: A Quantitative Scoring System.* The Psychological Corporation, London, 1987.

[36] Bishop, D. V. M. (1983). T.R.O.G. Test for Reception of Grammar. Manchester, UK: Chapel Press.

[37] Goldman, R., & Fristoe, M. (1986). Goldman-Fristoe Test of Articulation. Circle Pines, MN: American Guidance Service.

[38] Roberts, J.E., Schaaf, J.M., Skinner, M., Wheeler, A., Hooper, S., Hatton, D.D., Bailey, D.B. Jr. (2005). Academic skills of boys with fragile X syndrome: profiles and predictors. *American Journal of Mental Retardation, 110,* 107-120.

[39] Scerif, G., Cornish, K., Wilding, J., Driver, J., & Karmiloff-Smith, A. (2004). Visual search in typically developing toddlers and toddlers with Fragile X or Williams syndrome. *Developmental Science, 7,* 116-130.

[40] Jakala, P., Hanninen, T., Ryynanen, M., Laakso, M., Partanen, K., Mannermaa, A., & Soininen H. (1997). Fragile-X: neuropsychological test performance, CGG triplet repeat lengths, and hippocampal volumes. Journal of Clinical Investigation, *100,* 331-338.

[41] Busca Safont-Tria N. (2001). Psychomotricity and fragile X syndrome. *Revista de Neurologìa, 33 Suppl.1,* S77-81.

[42] Hodapp, R. M., Leckman, J. F., Dykens, E. M., Sparrow, S. S., Zelinsky, D. G., & Ort, S. I. (1992). K-ABC profiles of children with fragile X Syndrome, Down Syndrome and non-specific mental retardation. *American Journal on Mental Retardation, 97*, 39-46.

[43] Turk, J., & Cornish, K. M. (1998). Face recognition and emotion perception in boys with fragile-X syndrome. *Journal of Intellectual Disability Research, 42*, 490-499.

[44] Kogan, C.S., Boutet, I., Cornish, K., Zangenehpour, S., Mullen, K.T., Holden, J.J., Der Kaloustian, V.M., Andermann, E., & Chaudhuri, A. (2004). Differential impact of the FMR1 gene on visual processing in fragile X syndrome. *Brain, 127,* 591-601.

[45] Kogan, C.S., Bertone, A., Cornish, K., Boutet, I., Der Kaloustian, V.M., Andermann, E., Faubert, J., & Chaudhuri, A. (2004) Integrative cortical dysfunction and pervasive motion perception deficit in fragile X syndrome. *Neurology, 63,* 1634-1639.

In: New Research on Fragile X Syndrome
Editor: Jeffrey J. Upner, pp. 37-64
ISBN: 1-60021-281-6

Chapter 3

COGNITIVE-BEHAVIORAL PHENOTYPE AND CLINICAL MANIFESTATIONS OF FRAGILE X SYNDROME

Artigas-Pallarès Josep
The Neuropediatric Unit of Sabadell Hospital. Corporació Sanitaria Parc Taulí,
Barcelona, Spain

ABSTRACT

Fragile X Syndrome (FXS) is the most frequent cause of hereditary mental retardation. The neuropsychological manifestations of FXS, besides mental retardation (MR), also include a wide range of symptoms, such as behavioural manifestations and cognitive deficit. On the other hand, symptomatology of FXS is not limited to manifestations in individuals with the full gene mutation, but must also include the eventual consequences in bearers of premutation. The aim of this study is to contribute to a better understanding of the syndrome with the following objectives: 1) assess the frequency of the most relevant clinical manifestations in FXS; 2) describe the cognitive behaviour phenotype of FXS in a sample of Spanish patients; 3) determine whether the phenotype found is specific to FXS, or whether it could also be related to MR or to Attention Deficit/Hyperactivity Disorder (ADH/D) inherent to the syndrome; and 4) provide information about the consequences of the premutation in cognitive and behavioural aspects.

In order to fulfil these objectives, we carried out the following studies: Study 1 (106 patients): based on a questionnaire about medical and behaviour problems in FXS. Study 2 (30 FXS patients and two control groups: 30 MR and 320 ADH/D): a comparison between behaviours in FXS and behaviours found in patients with MR of other types and in patients with ADH/D. Study 3 (7 patients): neurocognitive manifestations found in premutated patients.

Conclusions From the sample of patients with FXS, the clinical characteristics typical of the syndrome can be described, as well as their frequency. A clear prevalence of epilepsy (14%) can be observed, a figure lower than those recorded in most studies relative to this phenomenon. The comparison with the MR and ADH/D groups show that behavioural manifestations in FXS are inherent to the genetic mutation, as they cannot be explained on the base of MR and/or ADH/D, which is almost always present in FXS. The assessment of bearers of the premutation of FXS proves the presence of cognitive symptoms in premutated males. This information reinforces the genetic model according to which the clinical manifestations of FXS are distributed in a continuous way, from low-range premutation to high-range full mutation.

INTRODUCTION

Fragile X Syndrome (FXS) is the most frequent cause of hereditary mental retardation (MR). The neuropsychological manifestations of FXS, besides MR, include a wide range of symptoms, such as behavioral manifestations and cognitive deficit. On the other hand, expressivity of FXS is not limited to the symptoms present in the full gene mutation, but must also include the eventual consequences in carriers.

The aim of this chapter is to contribute to a better understanding of the syndrome with the following objectives: 1) assess the frequency of the most relevant clinical manifestations in FXS and the characteristics of the cognitive-behavioral phenotype of FXS in a sample of Spanish patients; 2) determine whether the phenotype found is specific to FXS, or whether it could also be related to MR or to attention deficit/hyperactivity disorder (AD/HD) inherent to the syndrome; 3) provide information about the consequences of the premutation in cognitive and behavioral aspects.

CONCEPT OF BEHAVIORAL PHENOTYPE. HISTORICAL ANTECEDENTS

Human behavior is determined by both genetic and environmental influences. Yet the importance given to each one greatly varies according to the scientific approach.

During the XIX century and at the beginning of the XX, behavioral alterations now tackled by psychiatry were interpreted according to biological factors. Thus in 1895, S. Freud published a book entitled "A Project for Scientific Psychology", in which the author stated: "The cognitive mechanisms of mental normal and abnormal phenomenons can be explained with a rigorous study of brain systems". Nevertheless, molecular genetics, functional neuroimage and the study of brain biochemistry had not yet been developed in Freud's times. Had Freud had access to these means, he may have advanced in developing the ideas that he put forward in his project of scientific psychology, and consequently would not have ventured into the intricate and dense field of psychoanalysis, as the only approach to psychopathology. Maybe then his reasoning would have approached the present concepts on behavioral phenotypes, as they tackle behavior from biology.

The suspicion of the existence of genetically determined behavioral patterns was first raised by Money in 1963. He described the visuo-spatial deficiency in Turner Syndrome [1] a confirmation still valid today. In 1971, Belmont described the behavioral characteristics of Down Syndrome, as specific to this syndrome. In 1976, Nyhan [2] pointed out the behavior of self-mutilation as specific to Lesh-Nyhan Syndrome. This author was the first to use the term behavioral phenotype, [3] and his description of "genetic" behavior maybe the only one accepted to this day. Nevertheless, from the sixties to the eighties, the importance of biological psychiatry was discredited and these findings were underrated or ignored. It was not until the nineties that the concept of behavioral phenotype was retrieved, together with the development for new research techniques and the collaboration of neurologists, pediatricians, geneticists, psychologists and patients' parents to the study of genetic diseases.

Several other conceptual approaches have recently been suggested to explain the behavioral phenotype. It was initially defined as behavior in the general sense of the term (cognitive aspects and social interaction) associated to a specific syndrome with genetic etiology, where the phenotype is undoubtedly a result of a subjacent lesion. [4] Harris offers a much wider vision when considering behavioral phenotype any behavioral disorder not acquired [5]. A definition that synthesizes different approaches to this topic was recently suggested by O'Brien [6]. According to this definition, "*The behavioural phenotype is a characteristic pattern of motor, cognitive, linguistic and social observations that is consistently associated with a biological disorder. In some cases, the behavioural phenotype can constitute a psychiatric disorder; in others behaviours which are not usually regarded as symptoms of psychiatric diagnoses may occur.*"

The fact that it is defined as a characteristic pattern implies that we will be in the presence of specific coexisting behaviors that as a whole, make up the characteristics typical of a specific syndrome. Nevertheless, a specific behaviour must hardly ever be considered pathognomic of a determined syndrome, but any behaviour can appear in different syndromes. This does not exclude the fact that some behavioral manifestations can be almost specific to a determined syndrome, as occurs with the stereotyped hand washing typical of girls affected with Rett syndrome.

Another aspect included in this definition is that the behavioral phenotype is based on motor, cognitive, linguistic and social observations that are consistently associated to a biological disorder. The key word in the definition is "*observations*". This word was used rather than "alterations", "anomalies" or any other term that indicates a dysfunction. It is thus clear that behavioral phenotypes are based on the observation of the natural environment, regardless of theoretical considerations or validated criteria. This means that a child is labeled as shy when his/her parents or the people surrounding this child "*observe*" such a characteristic.

The first approximations to behavioral phenotypes were exclusively made from genetically based illnesses. Yet behavioral phenotypes were related to a "*biological disorder*", which can be genetic but not exclusively so. One can therefore also speak of a behavioral phenotype in non-genetic entities such as the fetal alcohol syndrome.

There is no doubt that research focused on understanding the relationship between genes and behavior will enable us to decipher many enigmas so far hidden, about the keys of behavior of the human species. As the genes involved in specific illnesses are being identified, and we are deciphering their action, new techniques of study are being introduced, that allow us to get a better insight on the relationship between the functions of genes and behavior. This approach: gene-behavior, can also be studied the other way round: behavior-gene. This allows an approximation to the understanding of the genetic bases of wide-spectrum disorders such as, amongst others, autism, dyslexia, AD/HD and Tourette syndrome.

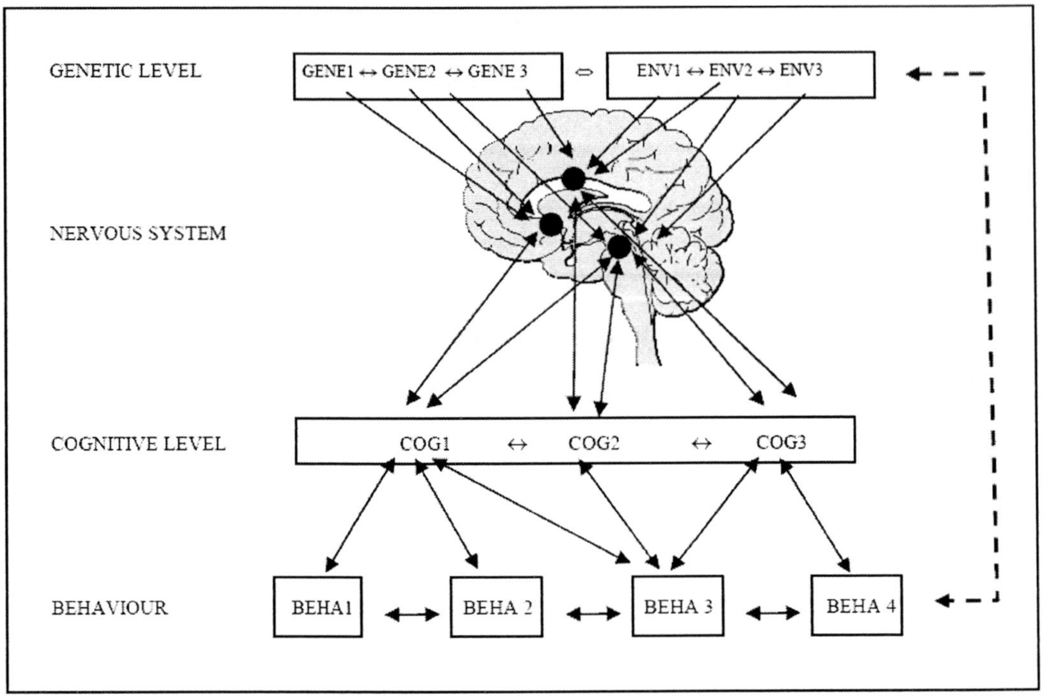

Figure 1. Levels of causality of behavioral phenotypes.

How Genes Influence Behavior

In order to determine a given behavior, the gene has an impact on the central nervous system at structural or biochemical level. Gene-modulated alterations affect the basic mechanisms of cognitive functions, which themselves will influence behaviors (figure 1).

Interaction mechanisms are extremely complex and occur at all levels.

The diagram above is a modification and amplification of the model suggested by Bishop [7] who was himself inspired by the works of Morton and Frith [8] and Pennington [9]

The most interesting aspects to be pointed out are:

1. Genes interact and are themselves modulated by environmental factors.
2. Both the brain circuits and the action of neurotransmitters are influenced by the genes and by the environment. Neurobiological structures are sometimes related to a single gene, although interaction between different genes and environmental factors is more common.
3. Behavioral patterns are related to cognitive mechanisms, determined by the operation of brain circuits. The repetition of behaviors influences the circuits, favoring or inhibiting certain behaviors.
4. Behavior modulates the environment in accordance with evocative influence and active influence.

Evocative influence means that the characteristics of the child determine the type of interaction with the people around him/her, which themselves revert on the child. We can take the example of an autistic child. The child's first interactions with his mother will undoubtedly be influenced by the impact of autism and his response to his mother's interactional needs will be poor. If the child does not respond to the mother's smile - this maternal behavior that facilitates the child's social development - it will be produced less often. Consequently, the autistic child will receive less social stimulus than a normal child.

Active influence means that a child, in accordance with his capacities, selects the activities that will facilitate or limit his cognitive development. For instance, a child with genetically based MR tends to look for fewer experiences likely to enrich his cognitive development. In other words, MR itself generates MR. These influences are an argument in favor of early intervention, which cannot modify the direct influence of the genes, related to the production of a protein, but can nevertheless limit both evocative and active genetic influence.

Another interesting aspect related to evocative and active genetic influence is the apparently paradoxical phenomenon that the influence of genes tends to increase with age. Traditionally, it was considered that genes influenced the first stages of life, and that behavior was later modified by the environment, separating it from genetic influence. Nevertheless, the opposite occurs, that is to say that the child resembles more his parents as he gets older. This is because the cognitive characteristics of the child are unstable during the first years, until they reach a certain stability in adulthood. This proves the relative low reliability of cognitive tests performed in the first years of life. On the other hand the modulation and selection of environments that will shape a definitive cognitive structure is influenced by the genetic characteristics of the individual. Furthermore, genetic influences tend to have long term influence, longer than environmental ones that can vary. Consequently, the accumulative effect of the genetic influence is present throughout the development.

How Can Behavioral Phenotypes Be Studied?

The study of behavioral phenotypes is complicated, not only for the enormous complexity of genetic mechanisms and interactions between genes, but also for the observation and assessment of behavior. Firstly, not all individuals with a specific syndrome have the same behavior. For instance, in FXS, which is one of the syndromes where behavioral phenotype seem best defined, there are severely autistic patients, totally disconnected from their social environment, who share the same diagnosis as FXS non-autistic patients. The same can apply to the intellectual level, that can vary from an intelligence within the limits of normality to serious retardation. A further difficulty is the lack of tools to assess specific behaviors, although they can be identified from simple observation. This is the case of language dysfunction and its pragmatic aspects. A third obstacle is that some syndromes have extremely complex genetic bases, as occurs in Rett Syndrome. Yet many aspects relating to gene functions and their interaction with other genetic mechanisms are still to be discovered.

Some syndromes present different genetic variants within the same syndrome. This is the case of Angelman Syndrome that has the following genetic variants: interstitial deletion (deletion of 15q11-q13 of maternal origin), mutation of "imprinting", uniparental disomy (both chromosomes 15 derive from the father), and microdeletion of gene UBE3A; or Prader Willi syndrome, with the same variants but in the paternal allele. In cases where different genetic forms of a syndrome exist, the assessment the phenotype-genotype relationship is of extremely great interest, as this knowledge will lead to a greater insight on the function of specific genes on behaviour.

Phenotypes can be described from well-defined mental disorders, based on accepted criteria, as with autism or AD/HD. But this approach does not allow for the differentiation of specific nuances for each syndrome. For instance, AD/HD is present both in Angelman syndrome and in Williams Syndrome, but the characteristics of AD/HD are highly different in both syndromes. This is why it is often necessary to refer to alterations that do not fit into accepted diagnostic classifications, as occurs with the language disorder of Williams Syndrome, or the self-aggressive behaviour of Smith-Magenis Syndrome. The former presents a pragmatic semantic disorder of language, with peculiarities typical to the syndrome. In Smith-Magenis Syndrome, the behavior of self-aggression is expressed in the very characteristic self-hugging.

The interpretation of behavioral phenotypes must not be exclusively limited to a static description of the symptoms, but consider that the evolution of these symptoms with time is also regulated by genes. One example easy to understand in relation to this is the evolution of the intelligence quotient (IQ). In normal conditions, the IQ tends to remain stable throughout life. However, this rule can be altered according to the genetic syndrome. In Prader Willi Syndrome and in Neurofibromatosis I, the IQ, still within the range of MR, tends to increase slightly with years. On the other hand, in Fragile X Syndrome and Williams Syndrome, the IQ decreases in adulthood, yet this does not imply the loss of the ability to acquire new knowledge.

Furthermore, there is a lack of tools to quantitatively and qualitatively assess behavioral aspects, especially if we take into account that some of the parameters we try to measure have not even been clinically defined. One example can be eye avoidance or tangential language, typical of FXS. These symptoms can easily remain unnoticed, as they have not been semiologically defined. But, although they may be observed, it is difficult to measure them quantitatively. In other cases, phenotypical aspects belong to folk knowledge rather than to scientific knowledge. Such is the case of stubbornness or tantrums.

If we take into consideration that this is a relatively new field, the approach to the knowledge of behavioral phenotypes, resorting to the tools at our disposition, has been very prolific.

Questionnaires are a very useful tool, as they allow the collection of the same information, in the same way, in a large group of patients. We now have at our disposition an index of questionnaires with different orientations. Some were designed to establish a diagnosis and the characteristics of the disorders in the autistic spectrum, such as the Australian Scale for Asperger Syndrome, the Childhood Autism Rating Scale (CARS), the Autism Diagnostic Interview-Revised (ADI-R) and the Autism Diagnostic Observation Schedule (ADOS). Other questionnaires are designed to the detection and diagnosis of various psychiatric disorders. Among them, the Child Behaviour Checklist of Achenbach (CBCL), the Conners Parent Rating Scale, the Conners Teacher Rating Scale and the Developmental Behaviour Checklist. Thirdly, there are also questionnaires aiming at the detection of problematic behaviors, such as: Aberrant Behavior Checklist, Society for the

Study of Behavioral Phenotypes Postal Questionnaire and the Vineland Adaptative Behavior Scales.

However, although some of the existing tools present good psychometric conditions, they cannot include very specific aspects for a particular syndrome. In order to avoid this difficulty, it is essential to design a questionnaire based on the characteristics of the syndrome. In order to prepare the questionnaire, we need to review the contribution of other studies on behavior and problems associated to it, so as to include all aspects typical of the syndrome.

In some cases, a control group is required. When assessing behavior, the control group must present MR to the same extent of the group studied, in order to be able to assess behavior really specific to a particular syndrome, and that are not simply related to MR.

Various logistic difficulties are considered in the study of behavioral phenotypes. On the one hand, one has to face the challenge of making a thorough study about very specific characteristics in rare disease. This entails that when undertaking research at this level, although latest technology is not required, one does need a clinical team that need be multidisciplinary, highly specialized and interested in behavioral aspects of one or various syndromes. An additional aspect is the geographical dispersion of the patients, a factor that limits the research team to reach an important amount of patients with the same syndrome. These patients have generally been diagnosed in their reference hospitals and do not require the application of sophisticated diagnostic or therapeutic techniques, (as far as medical assistance), it is therefore difficult to convince them to travel to the place where the study is being carried out. This problem can be solved with the creation of a support team, experts at a specific syndrome that can offer high-level support orientation. Nevertheless, this does not lessen the difficulty to gather a high "pool" of patients.

The conditions that allow getting round at least some difficulties are:

- Multicentric studies
- The use of questionnaires
- The collaboration of parents' associations.

Clinical Manifestations of Fragile X Syndrome

Clinical variability is related to various factors that include: [10]

1. Molecular variations in FMR1 that generate a range of levels of FMRP.
2. The combined effect of background genes that interact with FMRP directly or indirectly.
3. Environmental factors that can favor or prevent the development of the resulting dysfunction.

Some of the enigmas of the first aspect are presently just starting to be discovered, and the other two are still totally blurred to scientific knowledge.

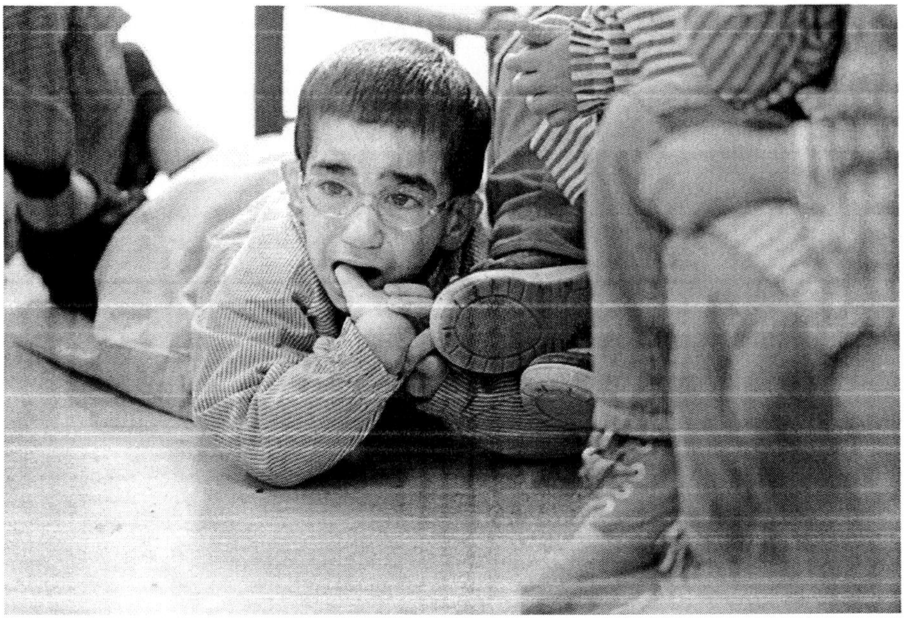

Figure 2. Physical characteristics.

1. Physical Aspect

Boys, and to a lesser extent, girls, present a peculiar physical aspect that can often suggest the syndrome. The most common features are a long face, large ears and a prominent chin. The facial physical characteristics of FXS are shown in Table I. Also see Figure 2.

It is important to bear in mind that almost one third of the patients do not present any identifiable physical sign. This emphasizes the importance of behavioral aspects, as a means of diagnostic suspicion.

Table 1. Facial Physical Characteristics

➢ Absent in 25 - 30 %
➢ Long face
➢ Prominent ears
➢ Prominent chin
➢ Prognatism
➢ Narrow palpebral fissures
➢ Macrocephalia
➢ Prominent forehead
➢ Epicantus
➢ Strabismus
➢ Hypotonia
➢ High arched palate
➢ Malocclusion of teeth
➢ Medical problems

The medical problems of FXS are partly related to connective tissue dysplasia and hypotony. Table II shows the various organs that can be affected, and the alterations most characteristic to a given syndrome for each organ.

Table II. Medical Problems of FXS

- Ophtalmological
  - Strabismus
  - Refractive errors
    - Myopia
    - Hyperopia
    - Astigmatism
  - Nystagmus
- Ear, nose, throat
  - Recurrent otitis media
  - Recurrent sinusitis
- Orthopedic
  - Pes planus
  - Joint laxity
  - Scoliosis
  - Articular luxations
- Skin
  - Soft and velvet-like skin
  - Sydney line
- Cardiocirculatory
  - Mitral prolapse
  - Aortic dilatation
  - Pulmonary artery dilatation
  - Hypertension
- Renal
  - Vesicouretral reflux
- Growth
  - Accelerated prepuberal growth
  - Slightly low stature in adulthood
- Endocrine
  - Postpuberal macroorchidism
  - Extreme obesity. Prader Willi-like
- Neurological
  - Hypotonia
  - Clumsiness
  - Convulsions
  - Insomnia

Table III. Behavioral Phenotype of FXS

➢ Lack of attention
➢ Hyperactivity
➢ Behavioral problems
➢ Shyness
➢ Social anxiety
➢ Poor eye contact
➢ Tactile defensiveness
➢ Tantrums
➢ Obsessiveness
➢ Repetitiveness
➢ Persistency
➢ Depression
➢ Aggressiveness
➢ Stereotypes
➢ Tics
➢ Pragmatic disorder of language. Tangential language

3. Cognitive-Behavioral Phenotype

The cognitive-behavioral phenotype of FXS is one of medium level MR. In individuals with the full mutation, mean IQ is 41. Patients with mosaicism (cells with the full mutation alternate with premutated cells) have an average IQ of 60. Those who have some cells with full methylation alternating with cells with partial methylation have an average IQ of 88 [11]. Within FXS, 13% of children do not have MR [12]. The K-ABC profile (Kaufman Evaluation Battery for Children), show in global scales, a sequential scale lower than the simultaneous one. The subtest with lowest score tends to be arithmetic. Thirty per cent of the children present a loss of IQ through the years, without losing abilities.

Undoubtedly, the most characteristic aspect is behavior. The physical features, as mentioned above, can be absent in one third of patients. The most interesting aspects are therefore MR and the peculiar behavior of the child, not merely as elements of diagnosis suspicion, but also as a valuable help to design behavioral and pedagogic intervention.

The most typical behavioral symptoms are shown in Table III.

Clinical Manifestation of the Premutation

Until recently, it was considered that the state of premutation was totally asymptomatic [13,14,15,16]. However, new data showing the opposite has progressively been presented. Although clinical manifestations in premutaded individuals can be subtle, they are worth taking into consideration, as they allow us to get a deeper insight into the disease.

The clinical effect in premutation is expressed at physical, cognitive, emotional and endocrinological levels, and in relation to the syndrome of progressive deterioration in old age. Premutated patients can present subtle features within the spectrum of physical features typical of FXS [17, 18]. Yet when looking for correlations between cognitive variables and level of FMRP, positive arguments in this sense started to emerge. [19, 20, 21] When studying more specific aspects of learning in premutated women, we found a lower performance in mathematical calculation in relation to reading and writing. This profile, found in premutated women, is the same, but less marked as that present in women with the full mutation. [22] Likewise, studies concentrating on specific neuropsychological aspects, such as executive function, have also found correlation in premutated individuals. [23]

Yet it is probably in the emotional aspect where the state of premutation is most clearly manifested. The most outstanding psychopathological features are social anxiety and emotional instability. This profile is also similar to the behavioral phenotype in full mutation. [24, 25] These problems cannot be explained by the psychological effect of having a child with a deficiency, as it has also been found in premutated individuals that do not have any child with FXS.

One manifestation clearly identified in premutated women is early menopause. [26] When screening was carried out for the diagnosis of premutation, it showed positive in 13 % - 16 % of women with familial premature ovarian failure and 3 % of women with non-familial premature ovarian failure. [27] In both cases, these figures are well above statistically expected. Sixteen per cent of premutated women start menopause before the age of 40, yet surprisingly, this does not occur in women with the full mutation. [28] The hormonal marker for menopause that was found to be altered is the follicle stimulating hormone. [29]

The most recent clinical findings among premutated individuals is the description of a syndrome of cerebellar tremor and cortical atrophy in elderly men carriers of the FMR1 premutation. Elderly men with the premutation and elevated levels of mRNA developed cerebellar tremor and cortical atrophy. They had previously shown difficulty to get dressed, drive, write and eat. MRI showed: dilated ventricles, slight cortical atrophy and severe cerebellar atrophy. The onset is between 53 – 63 years. Neuropsychological tests showed a loss of memory, as well as impairment of executive function and visuo-spatial abilities. The course is progressive. The conclusion to draw is that a moderate deficit of FMRP or an excessive increase of FMR1 mRNA can lead to retarded pathology of the central nervous system (CNS) in a limited number of premutated patients, possibly interacting with other mutations that affect CNS· [30, 31]

In premutated males, volumetric studies by MRI have also recently demonstrated significantly less voxel density in several brain regions, including the cerebellum, amygdalo-hippocampal complex and thalamus. [32] Furthermore, changes in flow and ínsula have also been demonstrated to be related to the level of mRNA. [33]

In contradiction to the classical model, the presence of a moderated decrease of the protein in the range of premutation has also been observed. According to the classical model, it was accepted that premutated individuals had a normal protein, as there was no methylation. Nevertheless, using the Willemsen immunocytochemical staining technique, a moderate diminution of the protein in the high range of premutation has been observed. [34]

The combination of information obtained with studies in premutated individuals and mRNA measurements in premutated and fully mutated individuals, has led to the suggestion of a new genetic model about fragile X syndrome, which modifies some of the traditional concepts. This model is represented in Figure 3.

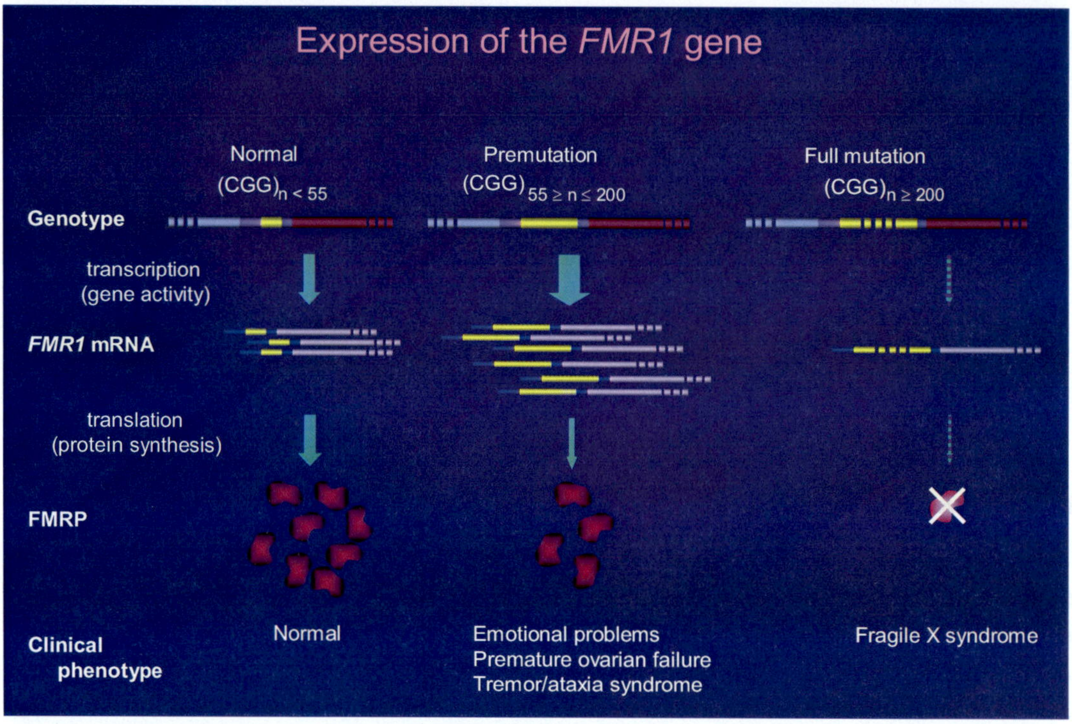

Figure 3. Influence of the number of CGG repetitions and the level of methylation in mechanisms of transcription and translation, according to P Hagerman. (Reproduction of chart with author's permission).

According to this model, besides the known deficiency in the transcription of the gene, there should also be a problem of translation. This was not taken into consideration previously, as it was unknown. In accordance with the classical model, there is a deficiency in the transcription derived from gene methylation; nevertheless, the finding of mRNA increase in the low premutation range implies the presence of an increase in transcription activity, maybe in order to compensate the deficiency in translation, that leads to a decrease of the protein.

Medical and Neurological Problems in Fragile X Syndrome Detected from a National Survey

This study aimed to obtain direct information on the clinical characteristics of FXS relating to medical and neurobehavioral problems, based on the information provided by the families of children affected by the syndrome.

In order to fulfill this objective, a questionnaire based on the medical and behavioral manifestations of FXS was sent to the families of 106 patients. In order to select items, the following books were revised: "Neurodevelopmental Disorders. Diagnosis and Treatment" (Hagerman RJ, 1999) and "Fragile X Syndrome. Diagnosis, Treatment and research. Third Edition" (Hagerman RJ, 2002), where all manifestations of FXS are described in detail.

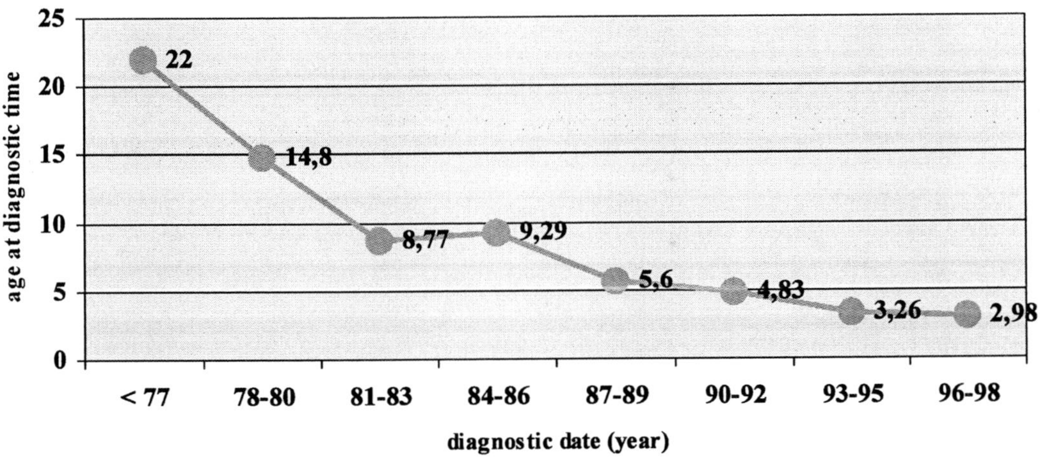

Figure 4. Evolution of age when diagnosis was made, according to year of diagnosis.

The questionnaire was sent by post by the Federación Nacional de Asociaciones del Síndrome X Frágil, with previously obtained consent. The National Federation sent the questionnaire to different regional associations to distribute to their local associates. The Asociación Catalana sent the questionnaire to families directly.

We revised 106 questionnaires corresponding to 95 boys and 11 girls, with ages ranging between 2.3 - 43 years, mean 12.79 (+/- 1.43) years.

1. Age of Diagnosis

Figure 4 shows a graph of the age at which diagnosis was made. Before 1977, the age of diagnosis was about 23 years. In the last few years (1996-98), mean age of diagnosis has been reduced to less than 3 years.

2. Development

The first manifestations of FXS are hypotonia with retardation of basic acquisitions.

The early appearance of these manifestations allow the physician to suspect FXS at about 2 years, or even before, if the degree of suspicion is high. Low muscular tone during the first year of life can be a sign alarm, observable during routine check up. It can be expressed as bad head control or joint laxity. The child is described as "soft".

Table IV shows a delay in initial walking and language, displayed in boys. The differences in these parameters are significative with regards to girls.

This information explains why the first reason for visiting the physician tend to be a delay in walking or language acquisition. Besides this, we also observed that 4% of the patients included in our study hardly manage to develop any form of spoken language.

Table IV. First steps and first words

	Girls	Boys	
First steps	14.8 months	19.2 months	$P < 0.05$
First words	20 months	33 months	$P < 0.01$

Table V Nose and hearing problems

	% in our study	% in other studies on FXS	% in normal population
Some chronic problem	54		
Recurrent otitis media	40	45 - 63	15
Recurrent sinusitis	4	23	
Insertion of PE tubes	17		

3. Ear, Nose and Throat Problems

FXS presents an alteration of connective tissue, which generated hearing problems as it facilitates recurrent otitis due to poor drainage of Eustachian tubes. The facial characteristics of the syndrome, long face and arched palate also favor infections due to the modification of the Eustachian tube angle, adding a further difficulty to the drainage of middle ear.

Over half of our sample presented some kind of chronic ear, nose, throat problems, the most frequent being recurrent otitis. Tympanic drainage of otitis serosa was required in 17% of patients. Table V shows the prevalence of such problems.

Infections can start during the first year and persist until 5 years. The symptoms are discomfort and fever. Sometimes, the child cannot express pain and expresses it with irritability and alterations of behavior. One of the most worrying consequences is loss of hearing.

4. Ophtalmological Problems

Patients with FXS often present visual problems. They may also be related with the alteration of connective tissue.

43% referred ophtalmological problems. The most frequent being strabismus, as shown in Table VI.

The conclusion drawn from this data is that all children with FXS must be referred for ophtalmological evaluation, as any sensorial problem added to the patient's cognitive deficiency will aggravate evolutive prognosis.

Table VI. Ophtalmological problems

	% in our study	% in other studies on FXS	% in normal population
Ophtalmological problems	43	30 - 56	
Strabismus	25	8 - 36	0.5 - 1
Refractive Problems	21	17	

Table VII. Orthopedic problems

	% in our study	% in other studies on FXS
Orthopedic problems	58	
Pes planus	52	50
Joint luxations	9	3
Hip luxation	5	
Hernias	7	15

5. Orthopedic Problems

One of the most typical signs of the syndrome, which is of great diagnostic interest, is metacarpophalangeal joint hyperextensibility. This is due to the fact that patients with FXS patients have hyperextensible joints due to hypotonia and loss of connective tissue. Microscopic studies of skin biopsies performed on FXS patients have shown a degradation of elastine fibers. [35]

Over 50% of patients present some kind of orthopedic problem, as is shown in Table VII. The most frequent orthopedic problem was flat feet, observed by patents in 52 % of cases.

Orthopedic problems must be tackled in order to avoid disorders derived from joint weakness. Depending on the specific problem, the use of boots, orthesis or physiotherapy may be indicated.

6. Sleep Disorders

Two types of problems with very different origins can alter sleep in FXS, insomnia and snoring.

In 14% of our sample, difficulties in getting to sleep were observed. This is due to an alteration of the hypothalamus, whose function is to regulate the production of melatonin. Compared to normal controls, children with FXS present lower levels of melatonin and an alteration of night secretion of this hormone. [36] The function of melatonin is to suitably maintain cycles of wakefulness and sleep. When this mechanism is not adequately regulated, difficulties in sleep onset or sleep maintenance throughout the night appear. In such cases, testing the administration of melatonin proves useful.

The other sleep disorder observed is a tendency to snore. This manifestation was registered in 45% of patients. Snoring is produced by hypotonia of bucopharyngeal muscles that causes a relative obstruction of respiratory tract during sleep, which reduces air circulation. In children who snore, it is of great interest to assess to what extent this affects their alertness during daytime wakefulness. Bad quality of sleep and lack of brain oxygenation can lead to aggravation of attention problems, as well as cause somnolence and bad mood during the day. All snorers must visit the otorhinolaringologist in order to assess possible problems of tonsillar or adenoid hypertrophy, which aggravate the situation and can be intervened upon. A more serious disorder related to breathing obstruction is sleep apneas. [37] In such cases, there is a more or less prolonged interruption in breathing, which may require the placing of an apnea monitor.

7. Neurological Manifestations

Neurological manifestations are very important in FXS as they determine the delay in development, behavioral aspects, cognitive alterations and the association with epilepsy. Alterations have been identified in different parts of the nervous system. Studies performed by functional neuroimage have supplied valuable information in this field. A reduction of the cerebellar vermis, thalamus and hippocampus [38] have been observed. Cerebellar deficiency seems to be responsible for the problems of sensory integration, motor coordination, and above all, of complex cognitive functions. A correlation has been found between the size of cerebellar vermis and performance on the following cognitive measures: total IQ, verbal IQ, manipulative IQ, cubes subtest (WISC-R), number of categories in Wisconsin Card Sorting Test and Rey Complex Figure. [39]

FXS present alteration in the bioelectric brain activity that can cause convulsions. This happened in 14% of the patients included in our study. This number is slightly different from those usually referred, which tend to be around 20%. [40] We think this difference may be due to the fact that our study is based on a population-based sample, and therefore lacks typical characteristics of hospital series that tend to include more serious cases. Nevertheless, another recent study on the prevalence of epilepsy in FXS refers 13.3% in children, a figure comparable to ours. [41]

Mean age for the first seizure was 5.4 years (+/- 2.1). Mean age when convulsions stopped was 7.7 years (+/- 2.8). Types of seizures varied widely, as is shown in Table VIII.

Epilepsy in FXS tends to be benign, given the relatively easy control of the seizures. They usually subside before the patients reach 20 years. Only in exceptional cases where FRM1 and FMR2 mutation coincide, can epilepsy be serious and difficult to control [42].

The evolution of the seizures is shown in Table IX

Table VIII. Types of seizures

➢ Generalized	8/15
➢ Partial simple	5/15
➢ Partial complex	2/15

Table IX. Evolution of seizures

➢ Average of crises during first year after initial crisis:	3.9
➢ No crisis alter first treatment	9/15
➢ No crisis alter some change	3/15
➢ Less than one crisis per month	2/15
➢ More than one crisis per month	1/15
➢ Still on one drug	4
➢ Still on more than one drug	2
➢ Does not take any medication	9

8. Behavioral Phenotype

Parents' observation of behavioral characteristics is shown in Table X.

The most outstanding problem is of attention deficit, with and estimated 85%. Attention deficit is practically constant in children with FXS, even in those with high performance. When studying this problem thoroughly with exhaustive neuropsychological tests, it was observed in 100% of patients, as was reported by Fryns in his detailed study including a sample of 21 boys. [43] Yet, although the attention problem is constant in FXS patients, they do not always fulfil the necessary criteria to be diagnosed of AD/HD. thus, Bregman in his study on 14 boys, appreciated that although 100% had attention problems, only 71% fulfilled the criteria of AD/HD. [44]

In FXS children, attention deficit has a particular cognitive profile, which differentiates them from other groups of patients who also present attention deficit. K Cornish [45] compared the characteristics of attention deficit between a group with FXS and a control group with Down syndrome. This study showed that children with FXS had more difficulty for the inhibition of repetition of irrelevant responses and for drawing attention in a sequence depending on whether the answer was correct. These cognitive mechanisms are executive functions, which is where the neuropsychological dysfunction of children with FXS lies. In women carriers of the full mutation, executive deficit has also been observed, although attention disorder is much less frequent [18].

Stereotypes also appear to be one of the most frequent symptoms in FXS, hand flapping being the most common (75%). Stereotypes in these patients depend on their state of anxiety. Children with FXS tend to present motor tics, complex tics, vocal tics, coprolalia, swearing, palilalia, as well as obsessive-compulsive symptoms. This suggests that some of the behavioral manifestations of FXS could be within the Tourette Syndrome spectrum. The association between these syndromes has been reported. [46]

Although not all children with FXS are hyperactive, hyperactivity is another of the most frequent symptoms, which appears in 73% of cases in our survey. The prevalence of hyperactivity is within the range described in other publications. According to the system used to classify a patient as hyperactive, the prevalence of hyperactivity has been reported to between 47 % [47] and 73 % [48] It must be pointed out that hyperactivity in FXS tends to improve with age. [49]

Table X. Behavioral characteristics observed by parents.

	% in our sample	% in other studies [1]
Attention problems	84	
Hand flapping	75	81 – 83
Hyperactivity	73	64 – 74
Overwhelmed by crowds	64	
Shyness	62	61 – 73
Overwhelmed by noise	61	
Social anxiety	59	
Poor interaction	57	
Stubbornness	56	
Hand biting	54	56 – 64
Obsessiveness	47	
Mutism	43	
Eye avoidance	40	86 – 98

Shyness is another of the most typical characteristics of FXS, especially in women with the full mutation, but also present in children. However, it tends to be less noticed in children, as it is overshadowed by AD/HD or by MR, which has been demonstrated to be more prevalent in boys than in girls. The frequency of shyness in our study is 62 %. Shyness is closely related to social anxiety, manifested in 59% of patients. Other related symptoms are poor visual contact (40 %) and mutism.

Sensory integration is the ability to process the information received by the senses. This function is affected in several ways in different clinical situations. [50] The symptoms directed related to sensory integration disorder (SID) result from the wrong information received from each of the senses and the interpretation and coordinated response between them. Children with FXS can have difficulties in processing visuo-spacial information and therefore feel very insecure when having to use the information about distances or spacial relations. Thus loss of balance often occurs. Individuals tend to under or overreact, thus increasing the problems. The child affected with FXS can feel overwhelmed when being caressed, rocked or hugged. He can also be extremely annoyed by a specific cloth texture, or the pressure a garment has on his body.

Auditory stimuli can also produce an unpleasant sensation that patients tend to avoid. Going to a supermarket, or being in any type of crowd can be an unbearable experience, as they receive an excess of stimulus: noise, movement, contact and images. Smells and tastes can also be a source of discomfort or avoidance. A sensation simply discomforting for other people, can become unbearable. The tendency of children with FXS to avoid eye contact can be related to SID problems, as visual stimulus proves too intense. In our study, 64% felt overwhelmed by crowd and 61% by noise. "Tantrums" (35 % of cases), hand biting, avoidance of eye contact, (40 %) and hand flapping (75 %), can be related to the difficulty FXS children have to process the stimuli that reach them. It has been suggested that hypersensitivity to stimuli can be related with an excessive interconnection or an excess of dentritic spines. [51]

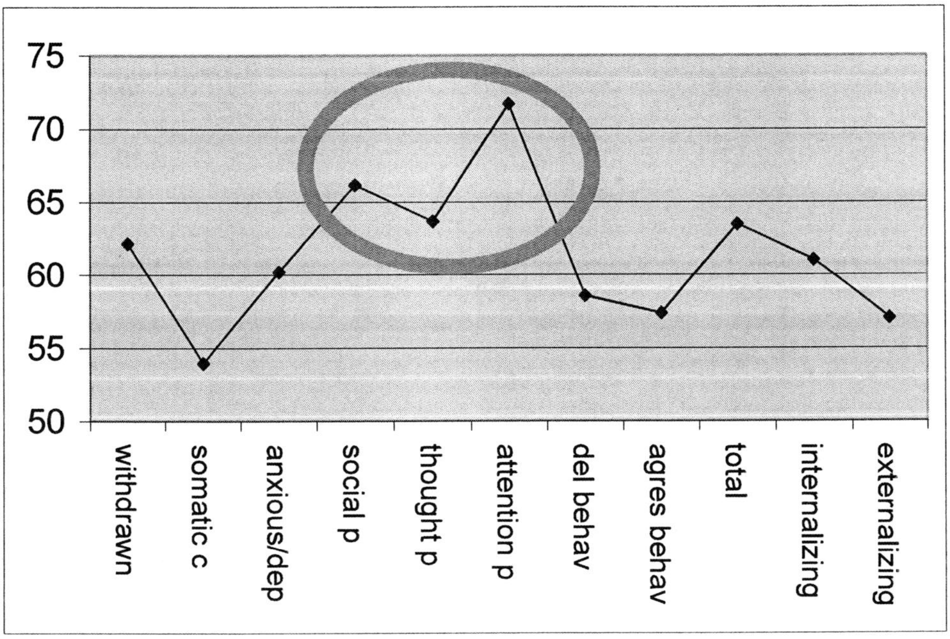

Figure 5. Scales scoring of CBCL 4-18 in FXS. Means of T scores.

The frequency of behavioral symptoms of FXS in our sample is generally very similar to that found in other groups, which reinforces the link between behavior and genetic defects of FXS. We approached our second objective with this speculation in mind.

Nevertheless, it could be argued that the behavioral symptoms of FXS are typical of children with MR and that in an unspecific way, coexisting MR and/or AD/HD is the cause of behavioral manifestations.

Table XI. Comparison of mean T scores of CBCL 4-18 scales between FXS group and MR group.

Scale	FXS. Mean score	MR Mean score	T student test
Isolation	62.15	62.23	NS
Somatic complaints	53.97	56.48	NS
Anxiety/depression	60.21	59.10	NS
Social problems	66.12	67.84	NS
Thought problems	63.67	63.94	NS
Attention problems	71.70	70.71	NS
Delinquent behavior	58.58	60.00	NS
Aggressive behavior	57.39	63.23	0.01
Total	63.45	65.81	NS
Internalizing	61.06	60.74	NS
Externalizing	57.09	61.94	0.01

CAN THE FRAGILE-X SYNDROME BEHAVIORAL PHENOTYPE BE EXPLAINED BY MENTAL RETARDATION OR ATTENTION DEFICIT/HYPERACTIVITY DISORDER?

This study aimed to establish the existence of behavioral aspects specific to FXS that are not related to MR or to AD/HD. Thus, we compared the behavior of a group of patients with FXS with two control groups: 1) a control group of patients with MR of different etiologies (MR group); and 2) a control group of patients AD/HD (AD/HD group).

The patients who participated in this study came from the neuropediatric outpatients' clinic of the Sabadell Hospital and from Xaloc Sabadell School - a special school for children with MR.

The MR control group was composed of 30 patients. The AD/HD control group was formed of 323 patients diagnosed of AD/HD. Age in the three groups ranged between 5 and 18 years.

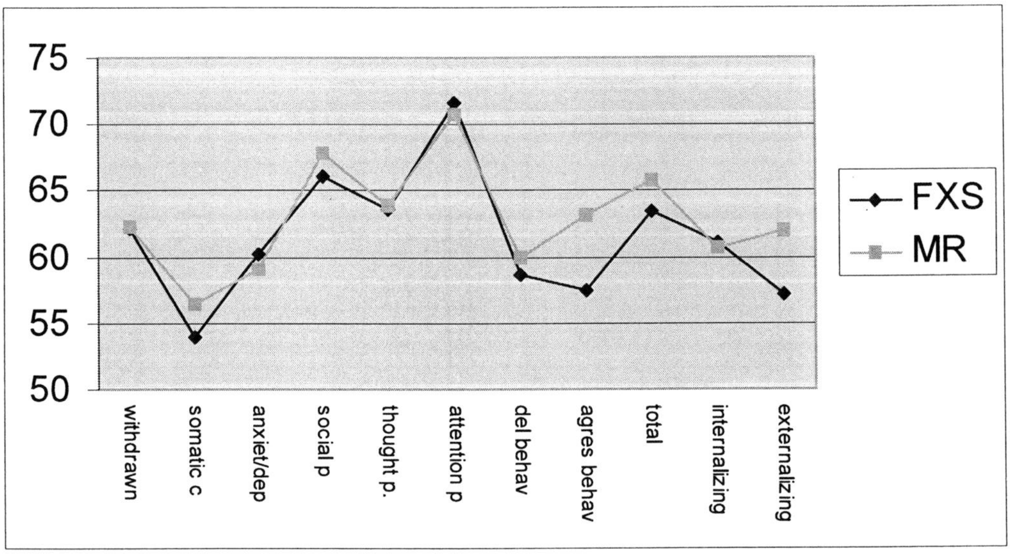

Figure 6. Comparison of CBCL 4-18 scales between MR group and FXS group. Mean T scores.

In the FXS and MR groups, the Stanford-Binet (Terman-Merrill, L-M 1975) test was administered. Behavior assessment was measured with the CBCL 4-18 Achenbach Checklist, 1991 Revision. The questionnaire includes 113 questions about specific behaviors. A Spanish translation previously used by the Outpatients' Pediatrics Section of the Spanish Association of Pediatrics was used. [52]

The assessment of CBCL 4-18 by scales, demonstrated that the scales with higher score in the FXS group were: social problems, thought problems and attention problems, as is shown in Figure 5.

The results of the scales show the most relevant psychopathological characteristics. scales that scored higher were: attention problems, social problems and thought problems. This reinforces the very high frequency of attention problems mentioned above.

In a study very similar to ours, but without a control group, Sullivan and cols. in North Carolina reached identical results. This study included a sample of 59 children between 4 and 12 years old. It showed that 51% of the children scored significantly in the total scale. Higher scores were obtained in attention, thought and social problems. [53]

In the comparison of scale results between the FXS and RM groups, patients with RM proved more aggressive with a higher score in the externalizing scale (figure 6). These results are in accordance with the impression that children with FXS, apart from some exceptions, tend to be fairly pacific, with less behavioral problems than other children with other types of MR.

Another way to analyze CBCL 4-18 responses is to assess the frequency with which subjects answer affirmatively to each of the items in the test. This way, we can assess the most frequent behaviors in children with FXS. Table XII shows the result of this analysis.

Table XII. Behaviors most frequently answered affirmatively in FXS group *

	% with this behavior
➢ CAN'T CONCENTRATE	100
➢ NERVOUS.	94.3
➢ ACTS TOO YOUNG	94.3
➢ SELF-CONCIOUS	4.3
➢ SHY	85.7
➢ SPEECH PROBLEMS	80
➢ POOR SCHOOL WORK	80
➢ Too dependent	80
➢ CAN'T SIT STILL	77.1
➢ REPEATS ACTIONS	77.1
➢ CLUMSY	77.1
➢ Swearing	77.1
➢ Demands attention	74.3
➢ Sulks a lot	74.3
➢ Acts without thinking	74.3
➢ Hot temper	74.3
➢ SHOWS OFF	74.3
➢ Fearful, anxious	71.4
➢ NERVOUS MOVEMENTS	68.6
➢ Screams a lot	68.6

* answered "sometimes" or "almost always" to corresponding item

In order to determine whether these behaviors, so frequent in the FXS group, could be characteristic of the syndrome, and therefore not related to MR or AD/HD, we compared mean punctuation for each behavioral item. In order to avoid the statistic bias derived from the comparison of multiple items we used the Bonferroni correction,. The results, that are shown in Table XIII, confirm the impression that some behaviors are highly related to FXS, as they are answered positively on the questionnaire significantly more than in MR and AD/HD groups. These results suggest that behaviors referred in this Table are more typical of FXS, as they could not be explained as an epiphenomenona of MR or AD/HD. In fact, this finding is in no way surprising, as it exactly fits with the behaviors that seem typical of FXS: self-confident, shy, obsessive, and speech problems.

The contribution of this study is the confirmation that some behavioral aspects are highly prevalent in FXS, and that these characteristics cannot be related exclusively to MR or AD/HD.

Clinical Manifestations of the Fragile X Premutation in Children

This study is based on the description of 7 boys, carriers of the mutation, who showed cognitive-behavioral symptoms. These patients were attended in a child neurology outpatients unit. All patients underwent:

- Hagerman Score
- Kaufman Assessment Battery for children (K-ABC)
- Number of CGG repeats
- FMRP level in 6 patients
- mRNA level in 6 patients

FMRP level was determined at Hospital Clínico in Barcelona by Dr. Montserrat Milà [54] mRNA level was determined at the M.I.N.D. Institute, University of California at Davis, by Dr. P. Hagerman.

Table XIV, XV, XVI and XVII shows respectively the age and diagnosis, physical characteristics, behavioral characteristics and the results of genetic testing.

Table XIII. Behaviors more frequent in FXS group than in RM and AD/HD groups (differences of means with Bonferroni correction)

Item	Mean differences with RM Group	Typical error	Significance	Mean differences with AD/HD Group	Typical error	Significance
Self-concious	0.99	0.16	0.000	0.79	0.12	0.000
Too dependent	0.69	0.18	0.000	0.44	0.13	0.003
Shy	0.61	0.17	0.001	0.80	0.13	0.000
Repeats actions	0.57	0.18	0.006	0.62	0.13	0.000
Speech problems	0.56	0.19	0.013	0.63	0.14	0.000
Fears	0.56	0.17	0.003	0.47	0.12	0.001
Concerned with neat, clean	0.42	0.15	0.018	0.36	0.11	0.004

Table XIV. Age and diagnosis of patients included in the sample

PATIENT	AGE	DIAGNOSIS
1	13	MR AND AD/HD
2	3	AUTISM
3	9	ASPERGER SYNDROME
4	9	AD/HD
5	11	LEARNING DISORDER
6	11	AD/HD
7	15	MR

Table XV. Physical Characteristics

CHARACTERISTIC	NUMBER OF PATIENTS
LONG FACE	5/7
LARGE EARS (>7 CM)	2/7
PROMINENT EARS	2/7
HIGH ARCHED PALATE	4/7
FINGER JOINTS LAXITY IN HYPEREXTENSION	4/7
HYPERKERATOSIS OF THE HAND	1/7
HEART MURMUR	1/7

Table XVI. Behavioral Characteristics

CHARACTERISTICS	NUMBER OF PATIENTS
HAND FLAPPING	3/7
HAND BITING	1/7
POOR VISUAL CONTACT	6/7
TACTILE DEFENSIVENESS	4/7
PERSEVERATIVE SPEECH AND BEHAVIOR	5/7
HYPERACTIVITY	6/7
ATTENTION DISORDER	6/7
TANTRUMS	4/7
SHYNESS	1/7

Table XVII. Genetic Features

PATIENT	REPETITIONS	FMRP LEVEL	mRNA LEVEL (CORRECTED)
1	53	< 28 %	2.05
2	53	> 70 %	1.23
3	60	> 70 %	1.56
4	50	> 70 %	1.52
5	50	NOT ASSESSED	NO ASSESSED
6	59	81 %	2.04
7	60	86 %	1.35

Few studies have been published on the premutation of FXS. In accordance with the classical model, it was considered that premutated individuals did not present any manifestations of the syndrome. Nevertheless, as was mentioned, manifestations relative to the following aspects have been described: physical, cognitive, emotional, endocrinological, besides a syndrome of progressive deterioration in advanced age.

In our study, we concentrated in premutated patients, within the low range of premutation, who presented some clinical manifestations. All patients presented specific symptoms that could suggest the possibility of FXS. This sample is therefore subject to bias. If the reason for requesting the test was the similarity with FXS, then the resulting sample obviously had to present characteristics very similar to those of FXS. We therefore cannot know whether these patients are really the most representative of the genetic anomaly studied, or whether they represent a subgroup.

The reasons for which our patients could have the manifestations found are explained by some of the following causes: 1) hiden expansions in tissues other than blood. Tissue mosaicism; 2) moderate deficit of FRMP; 3) moderate deficit of FRMP as a result of dysfunction in translation; and 4) Influence of other genes that cause cognitive or emotional deficits that have an additional effect to FRM1 premutation.

Rrecently, the association of premutation and disorders of autism spectrum have been reported [55, 56] In order to explain the association of fragile X premutation and autism, has been suggested the possible additional effect of a second gene. [57] The hypothesis of interaction between the premutation of FMR1 and other genes is highly interesting, as is permits understanding also the relations between FXS and autism. [58] Rogers and cols. compared the symptoms of a group of autistic children to the symptoms of a group of children with FXS and another group with developmental delay. They used ADI-R and ADOS-G – currently the two most accepted instruments for the diagnosis of autism. Surprisingly, they found that children with FXS could be defined in two groups, one that fulfilled the diagnostic criteria for autism according to the instruments used, and another group with did not fulfill the criteria for autism. As regards to symptoms, the FXS group with autism did not differ from the autistic group without FXS. The finding of two groups of FXS, one with autism and one without suggests that the association of FXS and autism is caused by the additional effect of other genes that act synergistically. [59]

No specific cognitive profile was found in our premutated patients, using the K-ABC, battery, either in global scales or for the subtest, which was to be expected if we take into consideration that the initial diagnoses were very diverse. With regard to behavioral characteristics, no typical behavioral profile of the premutation has been established, although numerous publications have appeared on this topic. The most prevalent symptoms observed in our patients were: poor eye contact, tactile defensiveness, preservative language and behavior, hyperactivity and attention deficit. This is in agreement with the observation of Dorn who, in 1994 carried out a study on carriers and control fathers. The premutated individuals showed a high incidence of behavioral problems, AD/HD, alcoholism, obsessive-compulsive symptoms, and difficulties in social interaction. [60]

REFERENCES

[1] Money J. Cytogenetic and psychosexual incongruities with a note on space-form blindness. *Am J Psychiat* 1963; 119: 820-7.

[2] Nyhan WL. Behavior in the Lesch-Nyhan syndrome. *J Autism Child Schi* 1976; 6:235-52.

[3] Nyhan WL. Behavioral phenotypes in organic genetic disease. Presidential address to the Society for Pediatric Research 1971. *Pediatr Res* 1972; 6:1-9.

[4] O'Brien G. Yule W. Why behavioural phenotypes?. En O'Brien G Yule W, eds. Behavioural phenotypes. *Clinics in Developmental Medicine* No. 138. London: Mac Keith Press; 1995. p. 1-23.

[5] Harris J C. Behavioural phenotypes in mental retardation: unlearned behaviours. *Advan Develop Disor* 1987; 1: 77-106.

[6] O'Brien G. The clinical relevance of behavioural phenotypes. En O'Brien G, eds. Behavioural phenotypes in clinical practice. *Clinics in Developmental Medicine* No. 157. London: Mac Keith Press; 2002. p. 1-12.

[7] Bishop D, Snowling MJ. Developmental Dyslexia and Specific Language Impairment: Same or Different?. *Psychological Bulletin* 2004, Vol. 130, No. 6, 858–86

[8] Morton, J., & Frith, U. (1995). Causal modeling: A structural approach to developmental psychopathology. In D. Cicchetti & D. J. Cohen (Eds.), *Developmental psychopathology* (Vol. 2, pp. 357–390). New York: Wiley.

[9] Pennington, B. F. (2002). *The development of psychopathology: Nature and nurture.* New York: Guilford Press.

[10] Hagerman RJ, Hagerman PJ. El síndrome X frágil: un modelo de la relación gen-cerebro-conducta.. *Rev Neurol* 2001; 33 (Supl 1): S 51-S57.

[11] Merenstein SA, Sobesky WE, Taylor AK, Riddle JE, Tran HX, Hagerman RJ. Molecular-clinical correlations in males with an expanded FMR1 mutation. *Am J Med Genet* 1996; 64: 388-94.

[12] Hagerman RJ, Hull CE, Safanda JF, Carpenter I, Staley LW, O'Connor RA, Seydel C, Mazzocco MM, Snow K, Thibodeau SN, et al. High functioning fragile X males: demonstration of an unmethylated fully expanded FMR-1 mutation associated with protein expression. *Am J Med Genet* 1994; 51: 298-308.

[13] Reiss AL, Freund L, Abrams MT, Boehm C, Kazazian H. Neurobehavioral effects of the fragile X premutation in adult women: a controlled study. *Am J Hum Genet* 1993; 52 : 884-94.

[14] Mazzocco MM, Pennington BF, Hagerman RJ. The neurocognitive phenotype of female carriers of fragile X: additional evidence for specificity. *J Dev Behav Pediatr* 1993; 14: 328-35.

[15] Taylor AK, Safanda JF, Fall MZ, Quince C, Lang KA, Hull CE, et al. Molecular predictors of cognitive involvement in female carriers of fragile X syndrome. *JAMA* 1994; 271: 507-14.

[16] Mazzocco MM, Holden JJ. Neuropsychological profiles of three sisters homozygous for the fragile X premutation. *Am J Med Genet* 1996; 64: 323-8.

[17] Riddle JE, Cheema A, Sobesky WE, Gardner SC, Taylor AK, Pennington BF, et al. Phenotypic involvement in females with the FMR1 gene mutation. *Am J Ment Retard* 1998, 102 : 590-601.

[18] McConkie Rosell A, Lachiewicz AM, Spiridigliozzi GA, Tarleton J, Schoenwald S, Phelan MC, et al. Evidence that methylation of the FMR-I locus is responsible for variable phenotypic expression of the fragile X syndrome. *Am J Hum Genet* 1993; 53: 800-9.

[19] Hagerman RJ, Staley LW, O'Conner R, Lugenbeel K, Nelson D, McLean SD, et al. Learning-disabled males with a fragile X CGG expansion in the upper premutation size range. *Pediatrics* 1996; 97: 122-6.

[20] Kaufmann WE, Abrams MT, Chen W, Reiss AL. Genotype, molecular phenotype, and cognitive phenotype: correlations in fragile X syndrome. *Am J Med Genet* 1999; 83: 286-95.

[21] Tassone F, Hagerman RJ, Taylor AK, Mills JB, Harris SW, Gane LW, et al. Clinical involvement and protein expression in individuals with the FMR1 premutation, *Am J Med Genet* 2000; 91:144-52.

[22] Lachiewicz AM, Dawson DV, McConkie-Rosell, Spiridigliozzi GA. Mathematics weakness in premutation females with the fragile X syndrome. 7 th International Fragile X Conference, July 2000, Los Angeles.

[23] Hills JL, Wilson R, Sobesky W, Harris SW, Grigsby J, Butler E, et alt. Executive functioning deficits in adult males with the fragile x premutation: An emergin phenotype. 7 th International Fragile X Conference, July 2000, Los Angeles.

[24] Franke P, Maier W, Hautzinger M, Weiffenbach O, Gänsicke M, Iwers B, et al. Fragile-X carrier females: evidence for a distinct psychopathological phenotype?. *Am J Med Genet* 1996; 64: 334-9.

[25] Franke P, Leboyer M, Gänsicke M, Weiffenbach O, Biancalana V, Cornillet_Lefebre P, et al. Genotype-phenotype relationship in female carriers of the premutation and full mutation of FMR-1. *Psychiat Res* 1998; 80: 113-27.

[26] Vianna_Morgante A M, Costa S S, Pares A S, Verreschi y T. FRAXA premutation associated with premature ovarian failure. *Am J Med Genet* 1996; 64: 373-5.

[27] Conway GS, Payne NN, Webb J, Murray A, Jacobs PA. Fragile X premutation screening in women with premature ovarian failure. *Hum Reprod* 1998; 13: 1184-7.

[28] Allingham_Hawkins D J, Babul_Hirji R, Chitayat D, Holden J J, Yang K T, Lee C, et al. Fragile X premutation is a significant risk factor for premature ovarian failure: the International Collaborative POF in Fragile X study--preliminary data *Am J Med Genet* 1999; 83: 322-5.

[29] Murray A, Webb J, MacSwiney F, Shipley EL, Morton NE, Conway GS. Serum concentrations of follicle stimulating hormone may predict premature ovarian failure in FRAXA premutation women. *Hum Reprod* 1999; 14: 1217-8.

[30] Hagerman RJ, Tassone F, Leehey M, Hills J, Wilson R, Landau W et al. Cerebellar Tremor and cerebellar cortical atrophy in older males with the fragile X premutation. 7 th International Fragile X Conference, July 2000, Los Angeles.

[31] Hagerman RJ. The physical and behavioral phenotype. En Fragile X Síndrome. Diagnosis, Treatment and Research. Third Edition. Hagerman RJ, Hagerman PJ, eds. Baltimore and London: The Johns Hopkins University Press; 2002. p. 3-110.

[32] Moore CJ, Daly EM, Tassone F, Tysoe C, Schmitz N, Ng V, Chitnis X et alt. The effect of pre-mutation of X chromosome CGG trinucleotide repeats on brain anatomy. *Brain* 2004; 127: 2672 - 81.

[33] Moore CJ , Daly ED, Tassone F, Jacobs P, Davis K, Murphy KC, et al. Neuroamnatomical effect of FMR1 gene mRNA in premutation carriers of Fragile-X syndrome. 9 th Annual Scientific Meeting. Society for the Study of Behavioural Phenotypes, November 2001, Oxford

[34] Kenneson A, Zhang F, Hagedorn CH, Warren ST. Reduced FMRP and increased FMR1 transcription is proportionally associated with CGG repeat number in intermediate-length and premutation carriers *Human Molecular Genetics*, 2001; 10: 1449-54.

[35] Waldstein G, Mierau G, Ahmad R, Thibodeau SN, Hagerman RJ, Caldwell S. Fragile X syndrome: skin elastin abnormalities. *Birth Defects* 1987; 23:103-14.

[36] O'Hare JP, O'Brien IA, Arendt J, Astley P, Ratcliffe W, Andrews H, et al. Does melatonin deficiency cause the enlarged genitalia of the fragile-X syndrome?. *Clin Endocrinol* 1986; 24:327-33.

[37] Tirosh E, Borochowitz Z. Sleep apnea in fragile X syndrome. *Am J Med Genet* 1992; 43: 124-7.

[38] Kates WR, Abrams MT, Kaufmann WE, Breiter SN, Reiss AL. Reliability and validity of MRI measurement of the amygdala and hippocampus in children with fragile X syndrome. *Psychiatry Res* 1997; 75: 31-48.

[39] Mostofsky SH, Mazzocco MM, Aakalu G, Warsofsky IS, Denckla MB, Reiss AL. Decreased cerebellar posterior vermis size in fragile X syndrome: correlation with neurocognitive performance. *Neurology* 1998; 50 : 121-30.

[40] Hagerman RJ. *Fragile X Síndrome. En Neurodevelopmental disorders. Diagnosis and Treatment.* Oxford University Press. New York: 1999. pp 61-132.

[41] Berry-Kravis E. Epilepsy in Fragile X Síndrome. 8 th International Frágile X Conference, July 2002. Chicago.

[42] Moore SJ, Strain L, Cole GF, Miedzybrodzka Z, Kelly KF, Dean JC. Fragile X syndrome with FMR1 and FMR2 deletion. *J Med Genet* 1999; 36: 565-6.

[43] Fryns JP. The fragile X syndrome. A study of 83 families. *Clin Genet* 1984; 26: 497-528.

[44] Bregman JD, Leckman JF, Ort SI. Fragile X syndrome: genetic predisposition to psychopathology. *J Autism Develop Disor* 1988; 18: 343-54.

[45] Cornish K, Munir F, Wilding J. Perfil neuropsicologico y conductual de los déficit de atención en el síndrome X frágil. *Rev Neurol* 2001; 33 (Supl 1): S 24-S29.

[46] Kerbeshian J, Burd L, Martsolf JT. Fragile X syndrome associated with Tourette symptomatology in a male with moderate mental retardation and autism. *J Dev Behav Pediatr* 1984; 5: 201-3.

[47] Finelli PF, Pueschel SM, Padre-Mendoza T, O'Brien MM. Neurological findings in patients with the fragile-X syndrome. *J Neurol Neurosurg Psychiatry* 1985; 48: 150-3.

[48] Hagerman RJ. Fragile X syndrome. *Curr Probl Pediatr* 1987; 17: 621-74.

[49] Borghgraef M, Fryns JP, Dielkens A, Pyck K, Van den Berghe H. Fragile (X) syndrome: a study of the psychological profile in 23 prepubertal patients. *Clin Genet* 1987; 32: 179-86.

[50] Ayres AJ. Characteristics of types of sensory integrative dysfunction. *Am J Occup Ther* 1971; 25: 329-34.

[51] Comery TA, Harris JB, Willems PJ, Oostra BA, Irwin SA, Weiler IJ, et al. Abnormal dendritic spines in fragile X knockout mice: maturation and pruning deficits. *Proc Natl Acad Sci U S A* 1997; 94 : 5401-4.

[52] Sección de Pediatría Extrahospitalaria de la Asociación Española de Pediatría. Estudio epidemiológico sobre la detección de la problemática psicosocial en el consultorio pediátrico. Monografía editada por Abbott Laboratoires, S.A. Madrid 1994.

[53] Sullivan K, Wheeler, Hatton DD, Baylkey DB. Behavioral characteristics of boys with fragile X Síndrome as assesed by the Child Behavior Checklist. 7 th International Fragile X Conference, July 2000. Los Angeles.

[54] M Rifé, J Mallolas, C Badenas, B Tazón, M Rodríguez Miguélez, T Pàmpols, A Sánchez, Milà M. Pilot study for the neonatal screening of fragile X síndrome. *Prenat Diagn* 2002; 22: 459-462

[55] [1] Farzin F, Perry H, Hessl D, Loesch DZ, Cohen J, et al. Autism spectrum disorders in boys with the fragile X permutation. *J Investig Med*, 2005; 53: S93-S94.

[56] Cornish K, Kogan C, Turk J, Manly T, James N, et al. The emerging fragile X premutation phenotype: Evidence from the domain of social cognition. *Brain Cogn* 2005; 57: 53-60.

[57] Goodlin-Jones BL, Tassone F, Gane LW, Hagerman RJ, et al. Autistic spectrum disorder and the fragile X permutation. *J Dev Behav Pediatr* 2004; 25: 392-8.

[58] Hagerman RJ. Lessons from fragile X regarding neurobiology, autism, and neurodegeneration. *J Dev Behav Pediatr* 2006; 27: 63-74.

[59] Rogers SJ, Wehner DE, Hagerman R. The behavioral phenotype in fragile X: symptoms of autism in very young children with fragile X syndrome, idiopathic autism, and other developmental disorders. *J Develop Behav Ped* 2001; 22: 409-17.

[60] Dorn MB, Mazzocco MM. Hagerman RJ. Behavioral and psychiatric disorders in adult male carriers of fragile X. *J Am Acad Child Adoles Psychiatry* 1994; 33: 256-64.

In: New Research on Fragile X Syndrome
Editor: Jeffrey J. Upner, pp. 65-76
ISBN: 1-60021-281-6

Chapter 4

THE COMPARISON OF PHENOTYPE IN MENTALLY RETARDED BOYS WITH AND WITHOUT FRAGILE X SYNDROME

Małgorzata Lisik
Silesian University School of Medicine, ul. Medyków 18, 40-752 Katowice, Poland

ABSTRACT

Fragile X syndrome is one of the most common hereditary forms of mental retardation, and has an estimated incidence of 1 in 4,000 males. The main characteristics of fragile X syndrome in males include mental retardation, macroorchidism, and peculiar face (elongated narrow face with large protruding ears and prominent jaw). In order to enlarge the number of variables useful in the preliminary evaluation of mentally retarded individuals, an analysis of data from medical records, physical examination, and anthropometric measurements have been undertaken. Forty two mentally retarded boys (age 4 to 20 years) with the characteristic features of fragile X syndrome, such as large protruding ears and elongated face were studied. Psychological examination including IQ and behavior disturbances was performed. Medical data, including birth weight were analyzed. Careful physical examination with special attention paid to minor malformations was performed. Anthropometric measurements were conducted in both groups. The measured variables included height, body weight, upper extremity length, lower extremity length, hand length, hand width, head length, head width, head breadth, total face height, bizygomatic diameter, bigonial diameter, nose length, nose width, ear height, ear width, inner canthal distance. Additionally, seven following indexes were calculated: cephalic, total face, mandibular-zygomatic, ear, nose, ratios of upper extremity length to stature high and lower extremity to stature high. Normative values for Polish children were obtained from published papers. Only in 16 boys methylated full mutations in the FMR1 gene were found, in 26 of them no recognizable cause of mental retardation was detected. The boys were divided into two groups: fragile X and non fragile X. High arched palate (44.44% v.s. 19.23%) and blue irises (83.33% v.s. 54.54%) were more often observed in FraX boys. The comparison between the average values of age adjusted anthropometric measures on all fragile X and non-fragile X boys showed significant differences with respect to height, head width, total face height and to the indexes of cephalic, ear and lower extremity length to stature. The greatest differences were found in ear width and weight. Two of the fragile X boys, 10 and 11 years old brothers with marked obesity (+ 4,89 SD and + 4,6 SD) were initially suspected to have Prader-Willi syndrome.

INTRODUCTION

The fragile X syndrome (FXS MIM 30950) is one of the most frequent hereditary forms of mental retardation, and has an estimated frequency of one in 4000 males. Fragile X syndrome is a very subtle dysmorphic syndrome and it is difficult to diagnose clinically. The characteristic features of fragile X syndrome are long face with prominent mandible, large and mildly dysmorphic ears and macroorchidism. This phenotype is subtle in young children and becomes more noticeable in adults. In young children the typical physical features are often not present. Macroorchidism has not yet developed and a long face is present mainly in adult males. Only large and protruding ears are the most common feature observed in young children with a full mutation in the FMR1 gene. The use of clinical criteria only especially in young children to recognize the fragile X syndrome is often difficult. The first clinical clue in children often is delayed attainment of one or more developmental milestones. Mental retardation in fragile X males varies from mild to profound with most affected males being moderately to severely retarded (IQ 20-55). A number of behavioral characteristics associated with FXS have been described. They include hyperactivity, short attention span, stereotypic behavior (hand–flapping, -biting, preservative speech), poor eye contact, tactile defensiveness and anxiety related to social contact. Additional findings may include mild connective tissue dysplasia, such as mitral valve prolapse, hyperextensible joints, and pes planus. The average age of diagnosis currently is 6-7 years, reflecting the subtlety of features in young children [9, 13, 14]. Due to the lack of clinical diagnostic criteria, simple scoring systems have been developed to select individuals for fragile X diagnosis, but these scoring systems are mostly useful for population-based studies.

The molecular basis of the fragile X syndrome has been confirmed by the isolation of the FMR 1 gene and CGG polymorphic trinucleotide repeat region in its 5' end. In the normal population the number of CGG repeats varies between 6 and 54. A full mutation is defined by a repeat length large then 200. The full mutation is associated with methylation of the CpG islands upstream of the FMR1 gene and with the subsequent shut down of gene transcription. As a consequence, no mRNA is produced and the lack of the gene product, FMRP, an RNA-binding protein, is responsible for clinical picture of the disease. Specific function of FMRP is not yet fully understood [13]. Unchanged FMR1 gene produces a protein that is thought to play a key role in both pre- and postnatal brain development. Fragile X Mental Retardation Protein (FMRP) is expressed in a variety of tissues, but it is most abundant in neurons. It has been hypothesized that FMRP may play role in mGlu-R-mediated plasticity, a process by which dendrites in the brain mature. People with the full mutation of FXS and lack of FMRP, have immature cerebral cortical spine morphology and increased dendritic spine densities. It, also, is postulated that FMRP by modulating mRNA translation is directly involved in synapse maturation during development and for synaptic activity in the adulthood [1].

Therefore, it is important for several reasons to diagnose children and adults with FXS. The proper diagnosis benefits the family because it ends their search for reasons why their child shows behavioral and cognitive deficits and it helps to focus on treatment and interventions that can benefit their child. Parents gain an understanding of FXS and how it modifies their child's development and behavior. Associated mental impairment and high risk of recurrence makes genetic counseling essential for families affected with FXS.

PATIENTS AND METHODS

Forty two mentally retarded boys age of 5 to 17 years with characteristic features of fragile X syndrome apparent by visual inspection, such as large and sometimes protruding ears, as well as elongated face have been examined.

Psychological Testing

Each boy was subjected to psychological testing using age-appropriate batteries. The Brunet-Lezine Intelligence Scale, the Termann-Merill Intelligence Scale and the Wechsler Intelligence Scale for Children-Revised (WISC-R) were used to test the intellectual abilities and presence of autistic features, behavioral disturbances. Also, IQs were determined.

Table 2 Results of psychological testing

Intellectual function		FraX boys		NonFraX boys	
	IQ scores	No	[%]	No	[%]
Below normal range	83-69	-	-	3	11
Mild degree mental retardation	68-52	7	44	7	27
Moderate degree mental retardation	51-36	6	37	1	42
Severe degree mental retardation	35-20	3	19	5	20
Profound degree mental retardation	19-1	-	-	-	-

Physical Examination

Careful physical examination with special attention paid to minor malformations was performed. All boys underwent a physical examination by one physician to document the presence or absence of physical features associated with the fragile X syndrome. The examiner subjectively evaluated the subject on the following features: long face, long ears, prominent ears, high-arched palate, scoliosis, hyperextensible finger joints, Simian palmar crease, iris color, thorax abnormalities.

Medical history of proband was also analyzed including data from the birth (birth weight) and developmental history (delayed motor skills, delayed speech and language development).

Table 1 Appearance of selected physical features typical of Fragile X syndrome

Feature	FraX group		NonFraX group	
	No	[%]	No	[%]
Long ears	16	100%	26	100 %
Prominent ears	16	100%	26	100 %
Long face	16	100%	26	100 %
Blue iris	15	83,33%	15	57,69 %
High arched palate	8	44,44%	5	19,23 %
Hyperextensible joints	7	38,89 %	9	34,62 %
Scoliosis	3	16,67 %	1	3,85 %
Epilepsy	2	11, 11%	3	11,54 %
Single palmar crease	1	5,56 %	0	0 %
Pectus excavatum	1	5,56 %	0	0 %

Anthropometric Measurements

Anthropometric measurements were made by one person according to the instructions and a set of measuring instruments manufactured by Martin and Saller [16]. The measured variables included 18 parameters: height, weight, upper extremity length, lower extremity length, hand length, hand width, head length, head width, head breadth, total face height, bizygomatic diameter, bigonial diameter, nose length, nose width, ear height, ear width, inner canthal distance and testes volume. Additionally, seven following indexes were calculated: cephalic, total face, mandibular-zygomatic, ear, nose, ratio of upper extremity length to stature and ratio of lower extremity to stature. Normative values for Polish children were obtained from publications. Z scores were computed using normative measurements to control age and sex, and were calculated from the formula $Z = (y-x)/SD$, where SD is the standard deviation of normal control individuals, x is the average of normal control individuals, and y is the measurement of the subject. Further, t-tests were used on the Z scores to determine if significant anthropometric differences exist between FraX and nonFraX boys.

Cytogenetic Testing

Chromosomal analysis followed the routine cytogenetic temporary cultures of blood lymphocytes in folic acid deficient medium (TC 199 and TC 199 with the addition of an inductor of folate-sensitive fragile sites: methotrexate for the last 24 hours of culture. The cultures were incubated for 96 hours at 37°C for the expression of the fra (Xq27.3) site. The cell divisions were arrested in methaphases by adding colcemid. The samples were fixed in a solution of 3:1 methanol-acetic acid, spread in slides. Manual analysis in light microscope included 100 methaphases from each culture scored and the presence of the fragile site at band q27 on the X chromosome (Xq27.3) was confirmed by GTG banding. Standard karyotype was made to exclude other chromosomal anomalies.

DNA Analysis

Genomic DNA was isolated from blood leucocytes by a standard salting-out procedure. DNA was double digested with the restriction enzymes EcoRI and the internal methylation sensitive enzyme BssH II. The resulting fragments were separated by gel electrophoresis in 1% agarose gel. Then samples were transferred into nylon membrane (Amersham) and hybridized with the DNA probe P2, which was an intragenic DNA, according to standard protocols. The radioactive products of hybridization were identified following autoradiography. The analysis of sizes of unstable CGG sequences was made according to method described by Rousseau et al. [15].

Pedigree Analysis

A comprehensive three generation pedigree from the consultands, parents of the proband, (e.g. offspring, siblings' offspring, parents, grandparents, aunts, uncles, nieces, nephews, and first cousins) were obtained using standardized pedigree symbols. Family history collection included questions about other family members, with particular attention to developmental delay, mental retardation, learning disabilities, autism, psychiatric disorders, attention deficit, and hyperactivity. Based on the results of the FMR1 gene analysis the subjects were categorized into the following two groups:

1. FMR1- positive group – FraX, with full mutation in the FMR 1 gene. The average age of FraX boys was 10.74, with a range of 5 to 17.
2. FMR1- negative group - non FraX, with normal allele of the FMR 1 gene. The average age of nonFraX boys was 10.35, with a range of 7 to 17.

Group comparisons were made between FraX and nonFraX boys.

RESULTS

Findings from Psychological Testing

In the FraX group IQ varied with age. In the group of preschool age 80% of boys revealed mildly mental retardation and in the remaining 20% the mental retardation was moderate. In contrast, in the school age boys (7 to14 years old) 33.3% were mildly mentally retarded, 44.4% showed moderate mental retardation and 22.3% revealed severe mental retardation. The decline in IQ value occurred with increasing age. Behavioral disturbances were also observed in FraX group. Attention deficit disorder was observed in all children in preschool age and in 77.7% of children at school age. Hyperactivity was observed in 60% of boys in preschool age group and 30% of boys in the group of school age. Hyperkinetic behavior was found to be age related, hyperkinesis was more frequent and intense in the preschool age group. No autism has been observed in the FraX group.

In the nonFraX group IQ varied. Twelve percent of boys showed IQ values below normal range, 27% were mildly, 42% were moderately and 19% were severely mentally retarded. Autistic like features were observed in 26.9% of examined boys.

Physical Examination

In the FraX group characteristic features for fragile X syndrome have observed such as: blue iris (83.33%), high arched palate (44.44%), hyperextensible joints (38.89%), scoliosis (16.67%), epilepsy (11.11%), single palmar crease (5.56%), pectus excavatum (5.56%) (Tab. 1). In the nonFraX group characteristic features for fragile X syndrome have been noticed such as: blue iris (57.69%) high arched palate (19.23%), hyperextensible joints (34.62 %), scoliosis (3.85%), epilepsy (11.54 %). There were no observations of single palmar crease and pectus excavatum (Tab. 1).

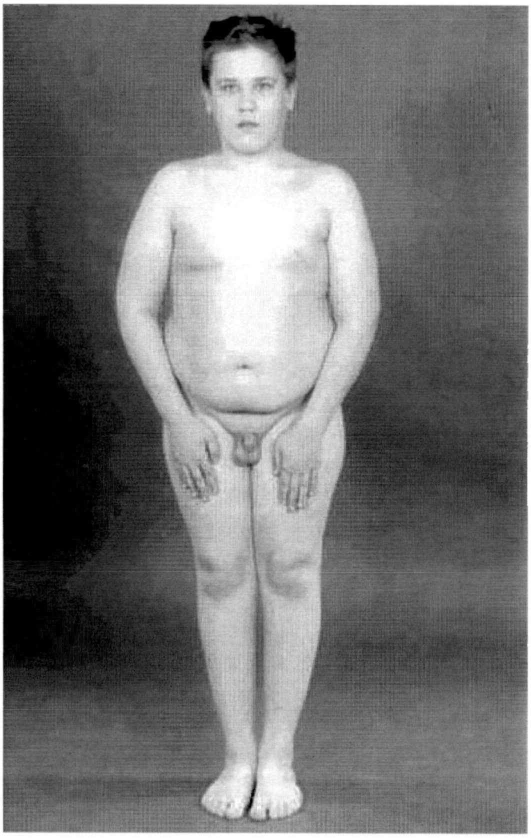

Figure 1. Photo of the 11-years old fragile X syndrome patient with Prader-Willi-like phenotype.

Anthropometric Measurements

The comparison between average values of age adjusted anthropometric measures on all Fragile X and nonFragile X boys showed significant differences with respect to height, head width, total face height and to the indexes of cephalic, ear and lower extremity length to stature. The greatest differences were found in ear width and weight. There were no significant differences between these two groups in other analyzed anthropometric variables. Macroorchidism was observed in all FraX boys older than 11 years and in 54.54 % of boys in non FraX group.

In the FraX group two brothers with unusual phenotype occurred. One was 10 and the other 11 years old. Both were obese with marked truncal obesity (S.D. respectively ± 4.89 and ± 4.6). Also both brothers did have a slightly elongated face, large, protruding ears, small hands and feet. However, their height was in the range of normal values. In the neonatal period there was no hypotonia and no need for nasogastric or parenteral feeding, the common characteristics of children affected with Prader-Willi syndrome, which they were initially suspected to have. The obesity was observed for the first time at the age of 4 to 5 years. The analysis of the FMR1 gene showed the presence of the full mutation in both of them. They were found to have 480 and 780 CGG repeats in the first exon of the FMR1 gene.

Cytogenetic Testing

Cytogenetic analysis revealed the presence of the fragile site fra(Xq27.7) in 13 boys in first method (TC 199) and 16 boys in second method (TC 199 with methotrexate addition). The mean frequency of cells with fra(X) (q27.3) was 7.75 % in TC199 Medium and 12.74 % after methotrexate induction. In a samples of 16 boys studied with both cell culture systems, a test for paired data showed that methotrexate was more efficient as an inductor of the fragile site than TC 199 medium alone. The standard karyotype analysis GTG bands did not show any other chromosomal abnormalities.

Molecular Analysis

Molecular studies using Southern blotting revealed the presence of abnormal size alleles (full mutation) in 16 boys, confirming the presence of mutations. All the patients have had full mutations in the FMR 1 gene. The number of CGG repeats ranged from 230 to 1,300. The presence of normal size alleles has been found in 26 boys.

All mothers of FraX boys were the carriers of premutation with the number of CGG repeats ranging from 60 to 160. They have been informed about their carrier status, the risk of recurrence of the disease in next children and availability of prenatal testing in future pregnancy.

Pedigree Analysis

Sixteen boys with confirmed fragile X syndrome belonged to 12 families. The data obtained from interview with parents on family histories were analyzed from obtained pedigrees and revealed that 75% of boys in FraX group had mentally retarded siblings and/or relatives, respectively. In the nonFraX group only 34.6 % of boys had mentally retarded siblings and/or relatives.

Discussion

Fragile X syndrome is the leading cause of familial forms of mental retardation. This is an X-linked inherited disorder. It is very important to diagnose affected patients as soon as possible to provide early intervention and supportive care (specific developmental therapy and an individualized education plan) and to inform parents for further parenthood planning. The phenotype in young children is subtle and becomes more noticeable in adults. In young children the typical physical features are often not present. Screening programs to facilitate the qualification to testing are needed. To increase the efficiency of screening programs, a preselection of clinical features is required. There are some reports, in which authors presented different fragile X checklists. Giangreco et al. [8] reported a useful six-item checklist based on physical features including: mental retardation, family history of mental retardation, large or protruding ears, elongated face, attention deficit hyperactivity disorder, and autistic-like behavior. Seven-item checklist published by de Vries et al. [20] included: family history of mental retardation, long face with high and wide forehead, large and protruding ears, hyperextension of the metacarpophalangeal joint digit V, soft and velvety skin, large testes (>30 ml), shyness personality, and lack of eye contact. Their data showed that individual clinical symptoms were not specific for the fragile X syndrome. It was rather combination of features, which considerably increased the likelihood of fragile X syndrome. A fragile X checklist proposed by Hagerman [9] included both physical and behavioral features. The typical physical features - prominent ears, long face, hyperextensible finger joints, flat feet, and macroorchidism - were helpful, but their absence was not unusual, especially in young children. Butler et al. [5] used a 15-item checklist, including physical and behavioral features frequently observed in fragile X syndrome in a prospective study of 188 mentally retarded males to identify males at risk for this syndrome. Nine out of 188 males were found to have fragile X syndrome. Six out of the 15 variables, plantar crease, simian crease, hyperflexibility, large testes, large ears and positive family history of mental retardation, were particularly good predictors of FraX syndrome. Limprasert et al. [11] tried to determine a cost-effective clinical checklist for fragile X syndrome screening in a Thai male pediatric population with developmental delay. They studied 179 nonfragile X male patients and 27 fragile X patients. They used six-item clinical checklist including family history, long and narrow face, prominent and large ears, attention deficit/hyperactivity, autistic-like behaviour and testicular volume. These were scored as 0 if absent, 1 if borderline and 2 if present. The best model was formula: 2 for family history of mental retardation + long and narrow face + 0,5 prominent and large ears +2 for attention deficit/hyperactivity + testicular volume = total score.

As in most clinical dismorphology studies, some clinical features might be differently scored in some patients by different observers. The items like family history, ears, joints and testes are objective measurements whereas the other features like face, skin and personality are rather subjective [20]. By using one examiner this variation maybe avoided. For less experienced clinicians a checklist would be helpful to assess whether a mentally retarded person should be tested for fragile X syndrome or not.

There are some specific, but not pathognomic, features associated with fragile X syndrome. These are not found in all patients and the frequency of each feature is age dependent. The most useful criteria in selection of patients to the diagnosis of fragile X syndrome found in this and other studies were: a family history of mental retardation, large or prominent ears, attention deficit hyperactivity.

The degree of mental retardation varied in both FraX and nonFraX boys. In agreement with observation made by Curfs [6] FraX boys in preschool age showed higher IQ than those older ones in school age. The IQ declined with the age. Attention deficit disorder in FraX group was a symptom observed in 100% of the preschool boys and in 77.7 % of school age boys. Borghraef et al. [2] noticed more common presence of attention deficit in children with FraX than in children with mental retardation of different etiology. They observed this symptom in 80% of preschool age boys and in 54% of school age boys. In the studied FraX group hyperactivity was observed more often in preschool boys (60%) than in school age boys (30%). Hyperactivity was found to be age dependent, more frequent and intense in the preschool age group. In contrast to intellectual performance behavioral disturbances decrease with age: the most striking of them hyperactivity and attention deficit was the most evident in young preschool FraX boys. Hyperactivity and attention deficit were observed in only 15.38% of nonFraX boys.

Some authors found association of fragile X syndrome with autism. Vieregge [18] observed autism in 13% of patients with FraX. The population analysis of patients with autism showed that 0 to 16% of them had FraX. This study did not confirm the previous observation and did not reveal association between FraX and autism. None of the FraX boys was diagnosed with autism. Autism was observed in 26.9% of nonFraX boys.

The anthropometric measurements were taken into consideration in the groups of fragile X patients. However, almost all of these studies were confined to adult males with a small number of children. These children were included in the general group. Butler et al. [3] undertook an anthropometric survey of 41 variables on mentally retarded males (39 with and 71 without FraX syndrome). The anthropometric variables consisted of: weight, height, sitting height, arm span, upper body/lower body segment ratio, total hand length, middle finger length, palm length, hand breadth, maximum fist breadth, maximum fist breadth, wrist breadth, ankle breadth, total foot length, head circumference, head length, head breadth, cephalic index, minimal frontal diameter, nose breadth, nose length, inner canthal distance, outer canthal distance, lower face height, face height, bizygomatic diameter, auricular head height, upper head height, head height, supraorbital diameter, mandibular arch, depth of lower third of face, philtrum length, ear length, ear width, shoulder-elbow length, elbow-finger-tip length, knee-buttock length, testicular volume. There was a statistically significant difference between FraX and nonFraX males with respect to 18 measurements: weight, height, 7 linear, 2 breadth, 5 craniofacial, subscapular skinfold and testicular volume, with the greater measurements found in the FraX males. Four anthropometric variables in the discriminant function were the Z scores representing testicular volume, ear width, head breadth, bizygomatic diameter. Butler et al. [4] developed anthropometric and craniofacial profile patterns on 34 physical parameters with 31 white mentally retarded males (23 adults and 8 children) with the fragile X syndrome matched for the age with 31 white, mentally retarded males without a known cause of their retardation. Significant differences were noted with greater head circumference, head breadth, lower face height, ear length and ear width in the fragile X syndrome males supporting an elongated face, large ears and macroorchidism in fragile X syndrome while bizygomatic diameter, inner canthal distance were significantly larger in the mentally retarded males without the fragile X syndrome. The comparison between average values of age adjusted anthropometric measures on all Fragile X and nonFragile X boys in our study showed significant differences with respect to height, head width, total face height and to the indexes of cephalic, ear and lower extremity length to stature. The greatest differences were found in ear width and weight. A significantly increased ear width was noted in the majority of FraX boys when compared with nonFraX ones. The weight of the investigated FraX boys was significantly higher than that in nonFraX boys [12].

Two boys in the FraX group, 10 and 11 years old brothers, had different phenotype from that typical for fragile X syndrome. They presented marked truncal obesity (+ 4.89 SD and + 4.6 SD), a slightly elongated face, large, protruding ears, small hands and feet. Their height was in the range of normal values. In the neonatal period there was no hypotonia and no need for nasogastric or parenteral feeding. The obesity was observed for the first time at the age between 4 and 5 years. They were initially suspected to have Prader-Willi syndrome. The "classical" phenotype of males with fragile X syndrome is that of a mentally retarded male with large testes and long face. There were other clinical stigmata seen in some FraX patients; obesity, small stubby hands and feet. Because of the clinical suggestion of Prader-Willi syndrome of these patients, the term of Prader-Willi-like phenotype was coined to describe the phenotype [17, 18]. There was not change in the diet. Fragile X syndrome should be included in the differential diagnosis of the obese child with mild to moderate developmental delay.

Fragile (Xq27.3) site could be observed in 2 to 60 % of metaphases, usually ranging from 4 to 16%. In some patients, very low expression (1 to 2%) or lack of fragile (Xq27.3) site does not exclude fragile X syndrome. These patients may have full mutation in the FMR 1 gene without the presence of fragile site (Xq27.3). In 3 of FraX boys presented in this study the fragile Xq27.3 fragile site was not detected in first method (low folate medium) and in only 1 to 2% in the second method (low folate medium with the addition of methotrexate). The method of choice in diagnosing FraX syndrome should be molecular analysis of the FMR 1 gene rather than cytogenetic methods due to false negative results.

Family history of mental retardation is an important risk factor for fragile X syndrome. This observation has been reported previously [9, 13, 14] and our data confirmed it. In investigated FraX group 75% of patients had affected siblings and/or relatives. Contrary only 34.6 % of nonFraX boys had mentally retarded sibling and/or relatives.

The diagnosis of fragile X syndrome can have far-reaching genetic and emotional implications for extended family members. Newly identified mutation carriers as well as families who have been previously diagnosed usually benefit from discussion of strategies or disclosing information about fragile X syndrome to other relatives, some of whom may react with anger, guilt, blame, disbelief, or indifference [13].

The carriers of premutation, especially males, have increased risk of a serious neurodegenerative disorders with progressive intention tremor and cerebellar ataxia (FXTAS-FragileX-associated tremor/ataxia syndrome). Affected person presents Parkinsonism, peripheral neuropathies, and dementia after the age of 50 years. Brain MRI studies of affected individuals are characterized by hyperintensities of the middle cerebellar peduncules. The penetrance of FXTAS appears to increase with age [10, 13]. Premature ovarian failure, defined as menopause occurring prior to the age of 40 may occur in up to 20% of female carriers of permutation, who should be informed about the potential for reduced fertility. Although surveillance for the clinical onset of POF is difficult, it may also be helpful to recommend close medical follow-up for early signs of POF. Discussion about these problems should be included during meeting with a clinical geneticist [9, 13].

Conclusion

The criteria needed for diagnosis of fragile X syndrome will depend on the age. Family history of mental retardation, large or prominent ears and attention deficit hyperactivity may be useful predictors in preselection of the mentally retarded patients for FraX syndrome testing.

The laboratory diagnosis should depend on molecular studies rather than cytogenetic methods.

Web Resource

Online Mendelian Inheritance in Man (OMIM) http://ncbi.nlm.nih.gov/Omim

References

[1] Bear M.F., Huber K.M., Warren S.T. (2004) The mGluR theory of fragile X mental retardation. *Trends in Neuroscience;* 27:370-377.

[2] Borghgraef M., Fryns J.P., Dielkens A., Pyck K., van den Berghe (1987) Fragile X syndrome: a study of the psychological profile in 23 prepubertal patients. *Clin Genet:* 32; 179-186

[3] Butler M.G., Alen G.A., Haynes J.L., Singh D.N., Watson M.J. Breg W.R. (1991) Antropomethric comparison of mentally retarded males with and without the fragile X syndrome. *Am J Med Genet:* 38;260-268

[4] Butler M.G., Pratesi R., Breg W.R., Singh D.N.(1993) Anthropometic and craniofacial patterns in mentally retarded males with emphasis on the fragile X syndrome. *Clin Genet*:44;129-38

[5] Butler M.G., Mangrum T.,Gupta R.,Singh D.N (1991) A 15-item checklist for screening mentally retarded. Males for fragile X syndrome. *Clin Genet:*39;347-54.

[6] Curfs L.M.G.,Wiegers A.M., Fryns J.P (1991) Intelligence and the fraX syndrome: a review. *Clin Genet:*2;55-62

[7] Fryns J.P.,Haspeslagh M., Dereymaeker A.M., Volcke Ph., van den Berghe (1987) A pecular subphenotype in the fra(X) syndrome: extreme obesity-short-stature-stubby hands and feet-diffuse hyperpigmentation. Further evidence of disturbed hypothalamic function in the fra(X) syndrome? *Clin Genet:* 32;388-92

[8] Giangreco C.A., Steele M.W, Aston C.E., Cummins J.H., Wenger S.L (1996) A simplified six-item checklist for screening for fragile X syndrome in pediatric population. *J Pediatr.:*129; 611-14

[9] Hagerman R.J. (2002) The physical and behavioral phenotype. In R.J.Hagerman & P.J.Hagerman *Fragile X syndrome, diagnosis, treatment and research* (3rd Ed. pp3-109) Baltomore: Johns Hopkins University Press.

[10] Hagerman R.J., Hagerman P.J. (2002) The fragile X premutation: into phenotypic fold. *Curr Opin Genet Develop:*12;278-83

[11] Lamprasert P.,Ruangdaraganon N.,Vasiknanonte P., Sura T.,Jaruratanasirikul S., Sriwongpanich N., Sriplung H (2000) A clinical checklist for fragile X syndrome: screening of Thai boys with developmental delay of unknown cause. *J Med Assoc Thai:* 83;1260-6

[12] Lisik M., Szymańska-Parkieta K., Gałecka U. (2000) The comparison of anthropometric variables in mentally retarded boys wit and without fragile X syndrome. *Clin Genet*; 57: 456-58

[13] McConkie-Rosell A., Finucane B., Cronister A.,Abrams L., Bennet R.L., Pettersen B.J. (2005) Genetic counseling for fragile X syndrome: Updated recommendations of the National Society of Genetic Counsellors. *Journal of Genetic Counselling*; 14:249-270.

[14] Nussbaum R.J., Ledbetter A The fragile X syndrome. In Scriver C.R., Beaudet A.L.,Sly W.,Valle D., Stanbury J.B. *The metabolic and molecular bases of inherited diseases.* New York; McHill; 1995:795-806.

[15] Rousseau F., Hitz D., Biacalana V., Blumenfeld S., Kretz C., Boue J., Croquette M.F., Gilgencrantz S.G., Jalbert P., Voelckel M.A., Oberle I., Mandel J.L. (1991) Direct diagnosis by DNA analysis of the fragile X syndrome of mental retardation. *N Eng J Med:*325;1673-82

[16] Martin R.,Saller K. Lehrbuch der antropologie. *Stuttgart: Gustav Verlag*; 1957:308-386

[17] Schrander-Stumpel C., Gerver W.J., Meyer H., Engelen J., Mulder H., Fryns J.P. (1994) Prader-Willi-like phenotype in fragile X syndrome. *Clin Genet*:45;175-80

[18] Vieriegge P., Froster-Iskenius (1989) Clinico-neurological investigations in the fra X form of mental retardation. *J Neurol*: 236;85-92

[19] De Vries B.B.A., Fryns J.P., Butler M.G., Canziani F., Wesby-van Sway E., van Hemel J.O., Oostra B.A., Halley D.J.J.(1993) Clinical and molecular studies in fragile X patients with a Prader-Willi phenotype. *J Med Genet*: 30;761-66.

[20] De Vries B.B.A., Mohkamsing S., van den Ouweland A.M.W, Mol E., Gelsema K., van Rijn M., Tibben A., Halley D.J.J., Duivenvoorden H.J., Oostra B.A., Niermeijer M.F (1999) Screening for fragile X syndrome among the mentally retarded: a clinical study. *J Med Genet*:36;467-70

In: New Research on Fragile X Syndrome
Editor: Jeffrey J. Upner, pp. 77-95
ISBN: 1-60021-281-6

Chapter 5

FRAGILE X AND STRESS: INSIGHT FROM STUDIES ON THE FRAGILE X MUTANT

Julie C. Lauterborn
University of California, Irvine, USA, CA 92697-4292

ABSTRACT

Fragile X syndrome is characterized by mental retardation and a behavioral phenotype that includes stress-related abnormalities in behavior. These include gaze aversion, hyperactivity, hyper-arousal, hyper-responsivity to sensory stimuli, impulsivity, social anxiety, and social withdrawal. Recent results indicate that the function of the hypothalamic-pituitary-adrenal axis, which is involved in the physiological and behavioral responses to stress, is altered in children with fragile X. Work in our laboratory has begun to investigate the stress response in a mouse model of fragile X, the Fmr1-knockout, to determine the extent to which the hypothalamic-pituitary-adrenal axis is dysfunctional and to identify potential cellular targets for therapeutics. Using a restraint stress paradigm, we find that, similar to the human condition, Fmr1-knockouts have an exaggerated glucocorticoid response to stress indicating perturbed negative feedback in this system. In addition, stress-induced changes in the expression of the immediate early gene c-fos within hippocampus, neocortex, and the hypothalamic paraventricular nucleus, as well as expression of brain-derived neurotrophic factor within hippocampus, are strikingly altered in the fragile X mouse. These results, as well as other findings to be described in this chapter, demonstrate that the fragile X mutants exhibit a greater response to stress than do wildtype mice, and support the hypothesis that the hypothalamic-pituitary-adrenal axis is dysregulated in fragile X syndrome.

INTRODUCTION

The hypothalamic-pituitary-adrenal (HPA) axis is involved in the regulation of the physiological and behavioral responses to stress through hypothalamic control over the secretion of adrenal glucocorticoid hormones and feedback mechanisms involving glucocorticoid modulation of activities in central neurons within the hypothalamus, pituitary, hippocampus, and frontal cortex. Disturbances in the HPA axis leading to chronic elevations in glucocorticoids (GCs) could have detrimental effects on forebrain neurons and, thus, profound effects on brain function. For example, increased levels of circulating glucocorticoids have been shown to exacerbate many types of neuronal insults, ranging from kainic acid lesions [1, 2] to ischemia-induced damage in hippocampus [3]. Chronically elevated glucocorticoid levels lead to hippocampal cell loss that is similar to what is observed in aging rats [4], whereas chronic stress leads to severe damage of CA1-CA3 hippocampal neurons in primates [5]. Restraint stress, which is accompanied by large elevations in circulating GCs, results in retraction of apical dendrites, a decrease in the number of dendritic branch points, and a loss of dendritic spines [6-8]. Yet, despite the deleterious effects of elevated GCs, this hormone is important for normal hippocampal function: too low levels of circulating glucocorticoids diminishes primed burst potentiation in hippocampal CA1 neurons [9] and the loss of glucocorticoids with complete adrenalectomy leads to apoptotic cell death of the hippocampal granule cells [10, 11].

There is recent evidence that the HPA axis is disturbed in fragile X syndrome and may be an important target for the management of the deleterious consequences of this genetic disorder. Fragile X syndrome (FraX) is the most common cause of inheritable mental retardation, affecting both males and females in an X-linked manner. The syndrome results from an unstable expansion of CGG trinucleotide repeats within the 5' untranslated region of the fragile X mental retardation 1 (Fmr1) gene. With >200 repeats, the expansion leads to transcriptional silencing resulting in diminished levels or total loss of fragile X mental retardation protein (FMRP). Current evidence indicates that FMRP acts as an mRNA binding protein that, in neurons, binds to its target mRNAs, and is involved in the transport of those mRNAs into dendritic spines. In the region of the spine, FMRP both inhibits translation under basal conditions but plays a role in neurotransmitter-activated protein translation (see [12, 13] for review). In addition to cognitive impairment and altered physical features, FraX is often characterized by abnormal behaviors including gaze aversion, hyperactivity, hyper-arousal, hyper-responsivity to sensory stimuli, impulsivity, social anxiety, and social withdrawal [14-20], behaviors that are often seen as "stress-related". This accords with work by Hessl and colleagues [21, 22] which indicates that the function of the HPA axis is altered in children with fragile X. Specifically, they find that male children with fragile X syndrome have higher salivary cortisol (the human form of GC) levels on normal, non-stressful days as compared to their unaffected siblings. On a "stressful" day, when they were presented with a social challenge and tested for cognition, the FraX children had greater than normal cortisol elevations, and greater severity of behavioral problems. These data were the first to suggest that the HPA axis is dysregulated in fragile X syndrome.

Our research is utilizing a murine model of fragile X, the Fmr1-knockout (Fmr1-KO) mouse, to investigate the stress-response in fragile X with specific interest in determining the degree to which the HPA axis may be dysfunctional, and if alterations in the stress-response play a underlying role in the cognitive impairment that is associated with this syndrome. The Fmr1-KO mouse was generated by insertion of a neomycin cassette into exon 5 of the murine *Fmr1* gene, resulting in animals that lack FMRP expression in all tissues [23]. To directly test the effect of stress in the Fmr1-KO (FX) mouse we have used a paradigm of restraint stress as previously described [24]. The paradigm entails restraining movement for measured periods of time (either 30 minutes or 2 hours) by placing them in 50 ml plastic centrifuge tubes (11.5

cm in length) with modified caps to accommodate the mouse's tail, and air-holes in the conical end of the tube. The tubes do not allow for forward, backward, or rotational movement thus effectively restraining the animal. Following a period of restraint, the mice are immediately sacrificed and brain tissues are processed to measure effects of stress on FX and paired wildtype (WT) mice. In particular, we are examining stress effects on gene expression for c-fos and brain-derived neurotrophic factor (BDNF) using radiolabeled *in situ* hybridization techniques. Both genes are regulated by neuronal activity as immediate-early genes [25, 26], and by stress [27-29]. Studies by others and us have shown that stress positively (c-fos) and negatively (BDNF and c-fos) regulates mRNA levels across numerous forebrain areas. Thus, analysis of these mRNAs provides a quantitative and topographic measure of the effect of the fragile X genotype on stress responses. Moreover, because BDNF is neuroprotective, studies of its expression can provide insight into mechanisms of neuronal vulnerability/damage. In addition to examining the effects of stress on gene expression, the corresponding changes in circulating stress hormones are evaluated. The following section describes the results of these studies to date.

Stress Effects on Glucocorticoid Levels in the Fragile X Mutant

Previous work form our laboratory has examined the effects of restraint stress on levels of circulating glucocorticoids in the Fragile X mouse [24]. Levels of corticosterone in blood serum were measured from Fmr1-KO and WT mice sacrificed without treatment (naïve controls) or promptly after periods of restraint stress lasting 30 minutes or 2 hours. Basal (control) corticosterone levels were not statistically different between genotypes, although there was slightly greater variability in the Fmr1-KO group than in the WT group. The reason for this variability is not clear but may indicate that under basal conditions the HPA is not tightly regulated in the FX mutant. This would accord with the observation by Hessl and colleagues [21] that under non-stressful conditions, FraX male children had higher salivary cortisol levels as compared to unaffected siblings.

Restraint stress significantly elevated corticosterone levels in both genotypes at both time points examined with the Fmr1-KOs exhibiting greater levels than the WTs (Figure 1A). The greatest difference between genotypes was observed at the 2 hour time point when levels were approximately 40% higher in the Fmr1-KO group ($p < 0.05$, for KO versus WT groups). Although this study did not examine the duration of increased corticosterone levels following the cessation of stress, it is noteworthy that corticosterone levels were further elevated from 30 minutes to 2 hours in the Fmr1-KOs whereas levels had reached a plateau at 30 minutes in the WTs. These results suggest that in the FX mutant there are disturbances in the negative feedback mechanism, which normally acts to limit the magnitude and duration of increased glucocorticoid levels, thereby resulting in greater circulating corticosterone levels in the FX genotype. In a preliminary report, Greenough and colleagues [30] describe similar results in both male and female Fmr1-KO mice using an acute stress paradigm. The finding of an exaggerated stress-induced glucocorticoid response in the FX mutant is again consistent with the study by Hessl et al [21] showing that stressors lead to greater increases in cortisol levels in FraX children as compared to unaffected siblings. Collectively, the FraX human and Fmr1-KO glucocorticoid data indicate that in this syndrome the hypothalamic-pituitary-adrenal axis may be in a primed state (i.e., may have elevated basal GC levels) under otherwise normal conditions, and hyper-responsive during stress due to a perturbed glucocorticoid negative feedback mechanism (Figure 2).

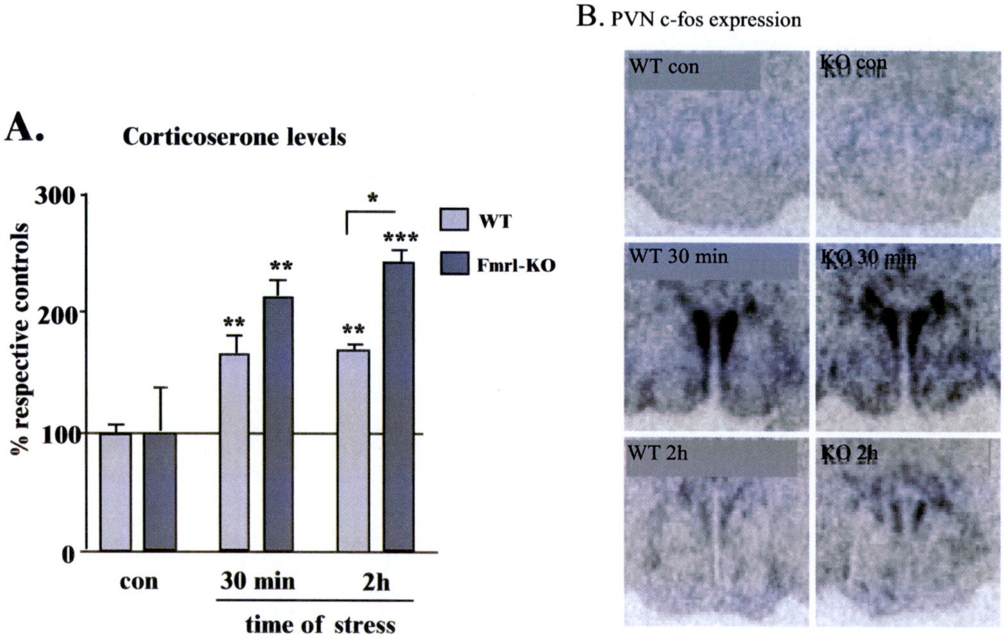

Figure 1. Stress-induced increases in corticosterone levels and hypothalamic gene expression are greater in the Fragile X mutant. Panel A shows quantification of blood serum corticosterone levels in naïve-control wildtype and Fmr1-KO mice, and following either 30 minutes or 2 hours of restraint stress; serum was analyzed by radioimmunoassay. Stress-induced increases in corticosterone levels were greater in Fmr1-KO mice (* $p < 0.05$, ** $p < 0.01$, *** $p < 0.001$ versus respective control group). Panel B shows the in situ hybridization localization of c-fos mRNA in the hypothalamic paraventricular nucleus (PVN) in naïve-control wildtype (WT) and Fmr1-KO (KO) mice (top row), and following either 30 minutes (middle row) or 2 hours (bottom row) of restraint stress. As shown, stress for 30 minutes lead to increases in c-fos mRNA levels in both genotypes whereas by 2 hours of stress elevations in expression were apparent only in the Fmr1-KOs.

How might the feedback mechanism become dysregulated in fragile X? The pituitary hormone corticotrophin (ACTH) is the principal regulator of cortisol/corticosterone synthesis and release from the adrenal cortex. Secretion of ACTH is regulated by the hypothalamic neuropeptide hormones corticotrophin releasing hormone (CRH) and arginine vasopressin (AVP). The CRH /AVP hypothalamic neurons receive input from multiple sources including the suprachiasmatic nucleus, brainstem nuclei, and cortico-limbic system which relay stress information.

The magnitude and duration of the HPA response is negatively regulated by GCs that act at both the pituitary and the hypothalamus to suppress synthesis and release of ACTH, CRH, and AVP (See Figure 2). The acute effects of GCs on ACTH and CRH/APV release are rapid, not requiring gene expression. In contrast, longer effects of GCs (lasting hours) occur by genomic mechanisms that involve de novo protein synthesis ([31] for review). Recent evidence suggests that annexin-1 (a.k.a, lipocortin-1), a member of the annexin superfamily of Ca2+ -and phospholipids-binding proteins that plays a role in the acute (non-genomic) glucocorticoid negative feedback mechanism [31, 32], may be dysfunctional in FraX. Annexin-1 is (i) expressed in the pituitary, hypothalamus, and hippocampus [33], (ii) up-regulated by GCs, and (iii) translocated from the cytoplasm to the surface membrane by a GC-dependent mechanism (see [31]). In leukocytes of FraX patients, annexin-1 is aberrantly modified post-translationally, which may result in abnormal function of this protein [34]. Although it is not clear if similarly aberrant processing of annexin-1 occurs in the pituitary or hypothalamus, these findings suggest that at least one mediator of the GC feedback mechanism may be dysfunctional in fragile X syndrome and warrants further exploration.

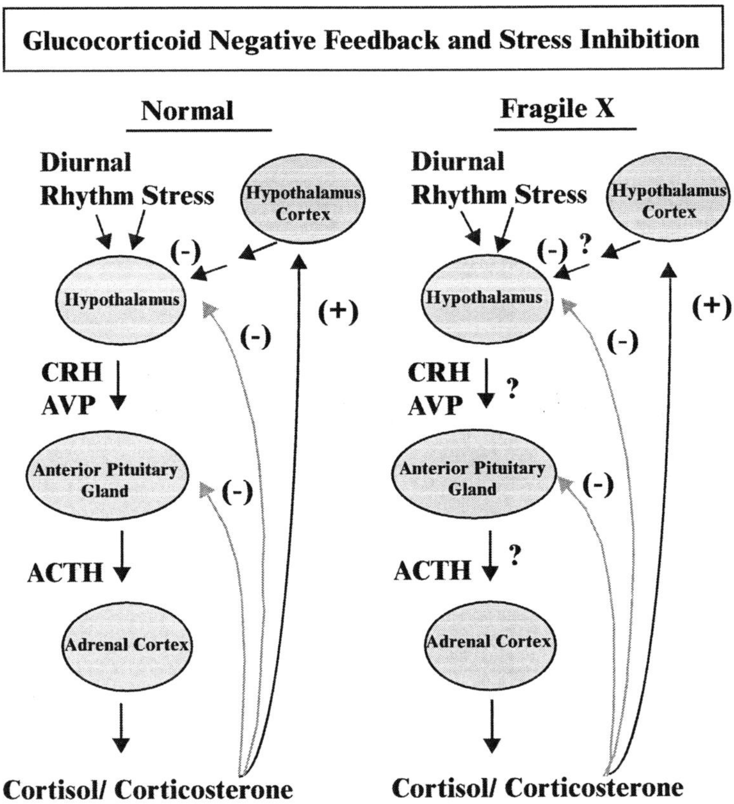

Figure 2. Schematic showing the HPA axis and the proposed dysregulated glucocorticoid negative feedback system in fragile X. Left panel shows the normal HPA axis and its response to stress and diurnal rhythms. The hypothalamic neuropeptide hormones corticotrophin releasing hormone (CRH) and arginine vasopressin (AVP) regulate the secretion of the pituitary hormone corticotrophin (ACTH) that, in turn, is the principal regulator of cortisol/corticosterone synthesis and release from the adrenal cortex. The magnitude and duration of the HPA response is negatively regulated by glucocorticoids (bold red lines) that feedback to act on the pituitary and hypothalamus to suppress synthesis and release of ACTH, CRH, and AVP. The glucocorticoid negative feedback system is complimented by a second cortical pathway, via glucocorticoid effects within hippocampus and medial frontal cortex, which acts to subsequently inhibit the hypothalamus. Right panel shows the HPA axis with the proposed altered feedback mechanisms in fragile X based upon findings in the Fmr1-KO mouse. As shown, the glucocorticoid negative feedback system is reduced (red lines), and the effects of glucocorticoids within hippocampus and cortex (i.e., changes in gene expression) are enhanced. Question marks indicate levels in the HPA axis, or in the feedback from hippocampus/cortex, where effects of the fragile X mutation have not been evaluated.

Finally, it is interesting that in the aged rat circulating basal corticosterone levels are slightly elevated [35, 36] and there is a reduced sensitivity to feedback regulation by corticosterone following stress [36, 37], similar to what is observed in fragile X. This has been interpreted as evidence that there is an age-related loss of sensitivity in the brain to the feedback inhibitory effects of high circulating corticosterone levels ([36] for review). In support of this view, Herman et al [38] has shown that in aged (30 month old) as compared to young adult F344/Brown Norway rats, both basal ACTH release and chronic stressed-induced CRH levels are elevated suggesting an overall increase in HPA tone with aging. Studies in our laboratory are currently investigating these other levels of the HPA axis (i.e., ACTH release and CRH levels) to determine if there is a generalized increase in HPA tone in FX mice that is similar to what has been reported for aged rats (Figure 2).

Stress and Gene Regulation in the Fragile X Mutant

Glucocorticoids signal via corticosteroid receptors that, once bound, translocate to the cell nucleus through an active transport mechanism involving a chaperone complex [39]. Once in the nucleus, receptor proteins dimerize, bind consensus DNA sequences, and either activate or repress gene transcription. The glucocorticoid receptor, one of two corticosterone receptors that have been characterized, binds glucocorticoids with low affinity and is extensively bound under conditions of adrenocortical drive such as stress [40]. Results described above, showing a greater corticosterone response with stress in the Fmr1-KOs, suggest there may be greater glucocorticoid receptor signaling in the cell nucleus and enhanced effects on gene expression in FX mutant mice.

To test this possibility we examined whether gene expression in the hypothalamic paraventricular nucleus (PVN), hippocampus, and neocortex is altered in the Fragile X mutant. In particular, we examined the effects of restraint stress on mRNA for the immediate-early activity regulated gene *c-fos* [24]. Messenger RNA for c-fos has been used in numerous studies to evaluate the neuronal response to stress, and is generally seen to be up-regulated in many forebrain fields including neo- and allocortical fields, lateral septum, hypothalamus and amygdala as a consequence of stress [29]. In addition, we have examined the effect of stress on the expression of brain-derived neurotrophic factor (BDNF), a gene known to be negatively regulated by glucocorticoids [27, 28, 41, 42], within hippocampus of the Fragile X mutant. The studies described here explored whether stress-induced changes in gene expression are more pronounced in the Fmr1-KO mice than in WTs.

Stress Effects on PVN c-fos mRNA Expression

To evaluate potential differences between Fmr1-KO and WT mice in the stress response within the neuroendocrine arm of the HPA axis we have focused on the hypothalamic PVN. As shown in Figure 1B, c-fos cRNA hybridization in the hypothalamus of control WTs and Fmr1-KOs was virtually undetectable. However, 30 minutes of restraint stress resulted in striking elevations in c-fos expression within the PVN of both Fmr1-KOs and WTs; the magnitude of this response was not different between genotypes (% of respective non-stress controls: 980 ± 187% for WT 30 minutes group versus 1085 ± 194% for Fmr1-KO 30 minutes group). These data are consistent with previous reports of the effects of restraint stress in rat and mouse that show peak c-fos mRNA expression levels at the 30 minute time point [29, 43]. However, after 2 hours of restraint stress differences between the genotypes were apparent. At this time point, c-fos mRNA levels had returned to near-control values for the WTs but were still significantly elevated in the Fmr1-KOs at nearly 600% of KO-control values. Overall, there was a significant effect of stress and an interaction between stress and genotype ($p < 0.0001$ and $p = 0.0294$, respectively, 2-way ANOVA) for c-fos expression within the PVN. These data reveal a more enduring hypothalamic response to stress in the FX mutants as compared to WT and are consistent with corticosterone measures (described above) demonstrating greater elevations in this hormone at the 2 hour time point in the Fmr1-KOs.

Stress Effects on Neocortical and Hippocampal c-fos mRNA Expression

Analyses of gene expression in the 30 minute and 2 hour stress treatment groups revealed further effects of genotype including different responses in hippocampus and neocortex [24]. With 30 minutes of restraint stress, there were marked increases in c-fos mRNA in the Fmr1-KO, but not WT, neocortex (Figure 3). As compared to naïve Fmr1-KO controls, increased expression in the Fmr1-KO stress group was observed throughout the neocortical mantle and was present in both the superficial (II/III) and deep (V/VI) pyramidal cell layers. Quantification of c-fos cRNA hybridization levels in the deeper layers at this time point demonstrated a 53% increase in c-fos expression in the Fmr1-KO group ($p < 0.001$ versus control group; Figure 3). In WT mice, labeling was somewhat greater in the 30 minute stress versus control groups but this effect was not statistically significant. It is notable, however, that in comparison to KO-specific changes within neocortex seen at 30 minutes, c-fos cRNA labeling was strikingly elevated in mid-line thalamic nuclei and hypothalamus of both genotypes at this time point (Figure 3). With 2 hours of restraint stress, neocortical c-fos mRNA levels were lower than in paired, non-stressed, controls for both genotypes. Overall, for the effect of 30 minutes and 2 hours restraint stress, 2-way ANOVA revealed significant effects of genotype ($p = 0.0423$) and stress ($p < 0.0001$) and a significant interaction between these two variables ($p = 0.0242$) within neocortex.

In hippocampus of control Fmr1-KO and WT mice, c-fos mRNA was similarly localized to the pyramidal cells of regions CA3-CA1, and to scattered cells distributed across other fields including the dentate gyrus granule cell layer, the hilus, and the molecular layers (Figure 4). Although basal levels of c-fos mRNA within the pyramidal cell layer were greater in control Fmr1-KOs than in control WTs (for overall effect of genotype, $p = 0.0254$, 2-way ANOVA), the main effect of restraint stress in both genotypes was a reduction in c-fos expression in this lamina [24]. The initial reduction in CA1 expression levels was greater in Fragile X mutant than in WT mice. Thus, after 30 minutes of stress c-fos mRNA levels in CA1 were reduced by 41% ($p < 0.01$) in Fmr1-KOs and by only 26% ($p > 0.05$, not significant) in the WTs. After 2 hours of constant restraint stress c-fos mRNA levels were comparably reduced in the two genotypes to 57% of respective control levels ($p < 0.01$ for both). Notably, this decrease resulted in an almost complete loss of c-fos mRNA within this cell layer for both genotypes (Figure 4).

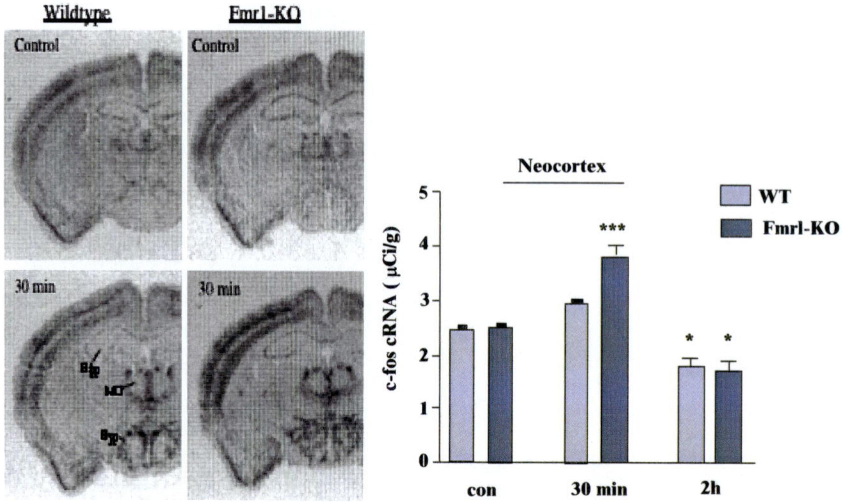

Figure 3. Stress-induced increases in neocortical c-fos expression are greater in the Fragile X mutant. Top panel show the in situ hybridization localization of c-fos mRNA in tissue sections through forebrain (dark-field photomicrographs) from naïve-control wildtype and Fmr1-KO mice (top row) and mice subjected to 30 minutes of restraint stress (bottom row). Stress-induced increases in cortical c-fos mRNA levels were principally localized to the superficial and deep layers of neocortex, with Fmr1-KOs exhibiting greater elevations than wildtypes. Stress-induced increases in labeling were observed within the mid-line thalamic nuclei in both genotypes, and appeared equivalent between the genotypes at this time point (hipp, hippocampus; Hyp, hypothalamus; MD, mediodorsal thalamic nucleus). Bottom panel shows quantification of c-fos mRNA labeling in neocortex following restraint stress. Stress-induced increases in c-fos expression at 30 minutes were significant only for the Fmr1-KO group. With 2 hours of stress, neocortical c-fos mRNA levels were equally reduced in both groups.

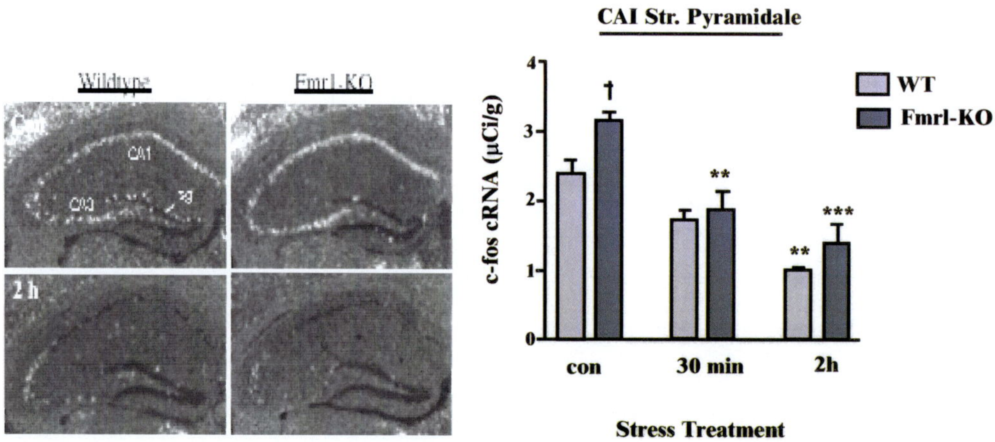

Figure 4. Stress-induced decreases in hippocampal c-fos expression in wildtype and Fmr1-KO mice. Top panel show the in situ hybridization localization of c-fos mRNA in sections through hippocampus (dark-field photomicrographs) from naïve-control wildtype and Fmr1-KO mice (top row) and mice subjected to 2 hours (bottom row) of restraint stress. As shown, stress greatly reduced c-fos cRNA labeling in the pyramidal cell layer of fields CA1-CA3 in both genotypes (sg, stratum granulosum). Bottom panel shows quantification of c-fos mRNA labeling in CA1 stratum pyramidale following restraint stress. Stress-induced decreases in c-fos expression were apparent by 30 minutes (significant only for the Fmr1-KO group; $**p < 0.01$ versus respective control group) and greatest by 2 hours (significant for both genotypes; $**p < 0.01$, $***p < 0.001$ versus respective control group).

In contrast to CA1, in the CA3 pyramidal cell layer the basal and post-stress c-fos mRNA levels were equivalent between Fmr1-KO and WT mice. Restraint stress reduced expression in both genotypes only at 2 hours (overall effect of stress: $p = 0.0011$, 2-way ANOVA; $p < 0.05$ for 2 hour groups versus respective controls). These data reinforce the impression that differences in the effect of stress on hippocampal c-fos expression between the FX mutant and wildtype are region-specific.

Stress Effects on BDNF Expression in the Fragile X Mutant

Brain-derived neurotrophic factor (BDNF) belongs to the family of neurotrophins, which also includes nerve growth factor, neurotrophin -3 and neurotrophin-4/5 [44]. The high-affinity BDNF receptor TrkB is broadly expressed in the brain [45] and, in accord with this, BDNF supports the survival of a variety of neuronal cell types including cholinergic, dopaminergic, and glutamatergic-cortical neurons ([46, 47] for review). Moreover, BDNF has been shown to affect dendritic spine morphology and density [48-50] and to play an integral role in synaptic plasticity including the facilitation of long term potentiation (LTP) ([51] for review), the probable cellular substrate of learning and memory. Thus, one would expect that suppression of BDNF expression and /or signaling could result in cognitive deficits and, in certain disease states, may be an underlying factor contributing to mental retardation.

Glucocorticoids have been shown to negatively regulate the expression of BDNF within hippocampus. In particular, restraint stress in rats reduces hippocampal BDNF expression [27, 28]. In addition, although adrenalectomy in rats has no effect on basal expression of BDNF mRNA, it does potentiate seizure-induced increases in hippocampal BDNF mRNA expression [41, 42]. Given this negative influence by GCs on hippocampal BDNF expression, we tested (i) if BDNF expression was similarly regulated by restraint stress in mouse and (ii) the prediction that the effect on BDNF expression would be more pronounced in the Fragile X mutant. Figure 5 shows that within hippocampus, restraint stress generally resulted in a greater reduction in BDNF mRNA levels within Fmr1-KOs than in WT mice. For the pyramidal cells of region CA1, 30 minutes of stress lead to a reduction in mRNA levels only in the Fmr1-KO group whereas levels in the WT-stress group were not different from naïve controls; differences in BDNF mRNA levels between the two genotypes were significant ($p = 0.035$). By 2 hours of stress, BDNF mRNA levels were reduced by 21% in the Fmr1-KOs but only by 7% in the WTs ($p < 0.001$ for Fmr1-KO versus WT groups); in this field there was a significant interaction between genotype and stress ($p = 0.0468$, 2-way ANOVA).

For the CA3 pyramidal cell layer and the dentate gyrus granule cell layer, analysis of BDNF mRNA levels by 2-way ANOVA revealed a significant effect of stress only ($p < 0.0001$ and $p = 0.0013$, respectively); there was no effect of genotype or any interaction between the two variables. Overall, these data demonstrate that the effects of restraint stress are more pronounced on hippocampal BDNF expression in the Fragile X mutant than in the wildtype mouse. As was found in the analysis of c-fos mRNA, effects were greatest for the early time point and for field CA1. A goal of future work in the FX mouse will be to determine if repeated exposure to stress (as we would anticipate occurs for Fragile X individuals in natural situations) leads to prolong decreases in BDNF expression and if such a loss in neurotrophic factor content has a detrimental effect on hippocampal synaptic plasticity.

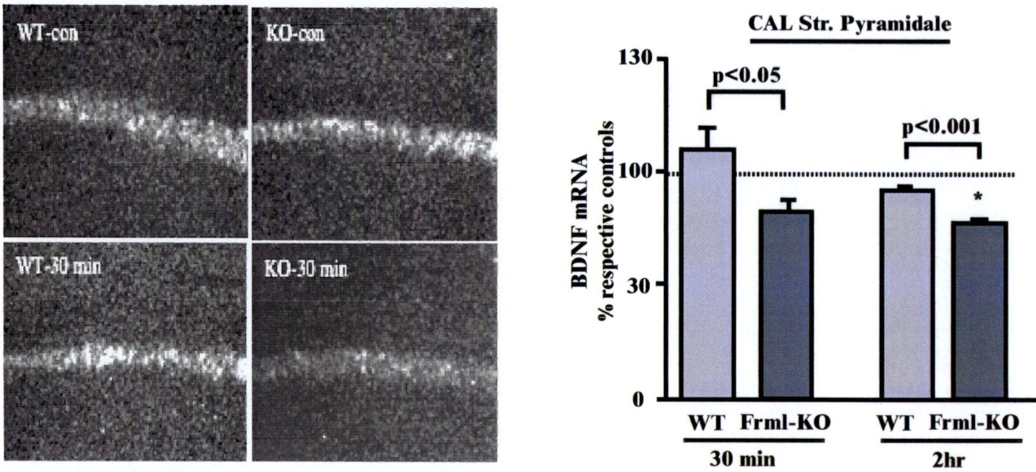

Figure 5. Stress-induced decreases in hippocampal BDNF expression are greater in the Fragile X mutant. Top panel shows the in situ hybridization localization of BDNF mRNA in the pyramidal cell layer of hippocampal region CA1 from naïve-control wildtype (WT) and Fmr1-KO mice (KO; top row) and mice subjected to 30 minutes (bottom row) of restraint stress. As shown, stress reduced BDNF mRNA in the Fmr1-KO with no effect in the WT. Bottom panel shows quantification of BDNF mRNA labeling in CA1 stratum pyramidale following restraint stress. Thirty minutes of stress reduced BDNF mRNA levels only in the Fmr1-KO group; at this time point levels were significantly different between the genotypes ($p = 0.035$). By 2 hours, mRNA levels were reduced in both groups although the decrease was greatest in the KO group ($p < 0.001$ for comparison between KO versus WT group; * $p < 0.05$ versus respective naïve-control KO group).

In summary, as predicted from the stress effects on corticosterone levels in the Fragile X mutant there are indeed greater effects on gene expression in the mutant (Figure 2). As discussed above, changes in gene expression following stress are mediated by GCs, but we have yet to determine if glucocorticoid receptor levels and signaling in Fmr1-KOs are altered following restraint stress. Recently, Ridder et al. [52] reported that glucocorticoid receptor-heterozygous mutant mice, producing only ~30% of normal receptor protein levels, have a disinhibited HPA system following stress similar to what we see in the FX mutant. This suggests that reductions in glucocorticoid receptor levels may play a role in the altered HPA response seen in FraX. The Fragile X mental retardation protein is believed to be involved in the suppression of dendritic translation [53-56] and recently it has been reported to associate with glucocorticoid receptor alpha mRNA [57]. Although under basal conditions total receptor protein levels are not altered in the Fmr1-KO, reduced glucocorticoid receptor immunoreactivity has been demonstrated in the initial dendritic segment of hippocampal region CA1 pyramidal neurons in the Fragile X mutant [57]; somal receptor protein levels were not analyzed in this study. This suggests (i) the novel view that glucocorticoid receptors may be translated in dendrites/spines, and indeed receptor protein has been localized to postsynaptic densities in amygdala [58], and (ii) glucocorticoid receptors in the Fmr1-KO are compartmentalized to the soma under basal conditions that may facilitate signaling and contribute to greater increases in gene expression. Given the pronounced effects of stress on c-fos within neocortex of the FX mouse, in addition to other reported abnormalities in dendritic spine shape and number [59] and reduced glutamate receptor levels [60, 61] within this field, we would predict that alterations in glucocorticoid receptor signaling are likely to be more pronounced within neocortex. Further work is needed to assess the degree to which levels, distributions, and signaling of the glucocorticoid receptor are perturbed in the FX mutant.

Anxiety-Related Behavior in the FX Mutant

The above results demonstrating that stress has greater effects on forebrain gene expression and circulating corticosterone levels in the FX as compared to WT mouse indicate that the long-term effects of a dysregulated HPA axis could have profound effects on behavior. Other laboratories have reported that FX mice exhibit anxiety-like responses in some behavioral tasks including the mirrored chamber test [62]. Yet, in other tests of anxiety including the open-field test, Fmr1-KOs do not exhibit increased anxiety-like behavior (i.e., more time spent in the periphery of the field and few excursions into the center of the field) ([62, 63]; Lauterborn, unpublished observations). In fact, Fmr1-KOs exhibit increased exploratory activity in the open-field test although this may be confounded by the observation that, in the same test, they have hyper-locomotor activity [62-65]. We have begun to examine the anxiety-behavioral phenotype of the FX mutant as well, and report, in this chapter, our first results.

The elevated T-maze test of anxiety has been extensively used to examine conditioned and unconditioned responses in fear/anxiety in rats [66]. The elevated apparatus consists of three arms of equal dimensions with one of these arms, perpendicular to the two opposed arms, enclosed by lateral walls. To examine conditioned anxiety (or learned fear/anxiety), animals are placed in the enclosed arm and the latencies to leave the arm and venture into the open arms are recorded over several trials. With acquisition of the conditioned fear response in this task, the latencies to leave the enclosed arm increase. Recently, Carvalho-Netto and Nunes-de Souza [67] described a modified version of the elevated T-maze test for use in mice that allows for reliable measures of learned fear responses in this rodent. In preliminary studies reported here, we used the modified apparatus which was (i) scaled down in size from that used for rats and (ii) included opaque (rather than transparent) walls to enclose the "start" arm of the maze (see [67] for details). Mice were tested on the apparatus for 5 consecutive trials with an inter-trial interval of 30 seconds. In each case, WT and Fmr1-KO mice (n = 10/ group) were individually placed at the distal end of the enclosed "start" arm and the time taken to exit the arm was recorded; a cut-off time of 300 seconds was employed for each trial. Although unconditioned fear (i.e., one-way escape from an open arm) can also be examined using this apparatus it was not tested in the present groups of animals.

As shown in Figure 6, there was a striking difference in the exit latencies between the two genotypes. For the WT group, the results are very similar to that reported by Carvalho-Netto and Nunes-de Souza [67]; the exit latencies for the first and last trials, and the overall acquisition profile (latencies across the trials), were similar between our studies indicating that they are highly reproducible results. In addition, we found a significant increase in the exit latencies in the WT group for Trials 4 and 5. By contrast, the Fmr1-KOs had significantly reduced exit latencies as compared to the WT group across most trials. These data indicate a marked deficit in the FX mutants for acquiring the task.

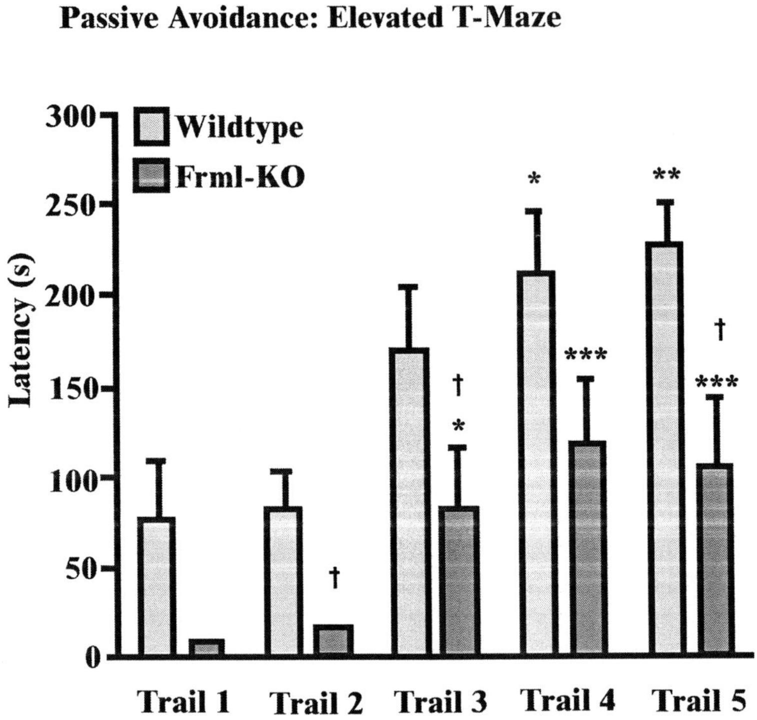

Figure 6. Fragile X mice exhibit altered behavior in a passive avoidance test. Bar graph shows latencies (seconds) to withdraw from the enclosed arm of an elevated T-maze. Avoidance was recorded over five consecutive trials (* $p < 0.05$, ** $p < 0.01$, ***$p < 0.001$ versus respective Trial 1 (baseline) for each group). The wildtype group (n = 10) exhibited increased latencies to leave the enclosed arm over Trials 3-5; this represents an instance of conditioned anxiety. In contrast, the Fmr1-KO group (n = 10) exhibited lower latencies in all trials (□ < 0.05, comparison between KO and WT groups for each trial).

In agreement with the present findings, deficits in conditioned fear responses in FX mutants have been reported by some (Paradee et al., 1999), whereas others have found no differences in between Fmr1-KOs and WTs [23, 68]. The seemingly conflicting results across laboratories may reflect differences in the testing paradigms used. The elevated t-maze test used to generate results described in this chapter, also has been used to show that the conditioned fear response is sensitive to hippocampal damage, with lesions restricted to the ventral, but not dorsal, hippocampus resulting in reduced exit latencies [69]. Although the Fmr1-KO does not exhibit gross pathology within hippocampus [70], differences in the distribution of the hippocampal mossy fibers in the FX mutant have been reported for the ventral hippocampus [65, 71]. It is interesting to consider that abnormal circuitry within the ventral hippocampus of the FX mutant may underlie impairment of the feared conditioned response.

CONCLUSION

Implications to Mental Retardation

As discussed above, the limited clinical and animal studies on stress in FraX indicate that there is a dysregulated stress response in this syndrome. Specifically, there appears to be a perturbed negative feedback mechanism resulting exaggerated responses to stress including excessive adrenal output and altered stress-induced gene responses. The degree to which the HPA axis is dysregulated and glucocorticoid receptors are abnormally localized within hippocampal neurons in FraX is critically important to understand because these abnormalities may play a role in mental retardation. There are two lines of evidence to suggest this possibility. First, it is clear that stress and/or anxiety leads to cognitive impairment in both animals and humans [72, 73]. The hippocampus is critical for declarative memory. It is also a primary CNS target for corticosteroids and provides negative feedback to the HPA axis. Numerous studies have shown that stress and high levels of corticosteroids damage hippocampal neurons, suppress hippocampal neurogenesis, and lead to memory deficits ([74] for review). Thus, further understanding of how stress affects the neurochemistries of hippocampus in the Fmr1 mutant brain is an important first step towards elucidating the influence of the HPA axis on mental retardation.

Second, it is broadly accepted that dendritic spines, and spine synapses, are the neuroanatomical loci of plastic changes with learning and memory [75, 76]. Dendritic spine abnormalities have been observed in FraX cerebral cortical autopsy material [77-80], and abnormal, immature-looking dendritic spines and greater spines densities have been reported on neocortical neurons in adult Fmr1-KOs [59], suggesting that these abnormalities may underlie cognitive deficits associated with this syndrome. Although excessive corticosterone levels and chronic stress have been well-described to result in dendritic loss, neuronal atrophy [6, 81] , and a rearrangement of dendritic arbors [82], other studies also have shown that more modest elevations in corticosterone in rats, and acute stress, lead to increased dendritic spine densities on cortical neurons [83-85]. These findings suggest that dysregulation of the HPA negative feedback mechanism and an exaggerated stress response in FraX could act to modulate dendritic spine density resulting in greater numbers. Future work in our laboratory is aimed at understanding the role of aberrant stress hormone levels in dendritic spine abnormalities in FraX and determining if these manifestations of disturbances in the HPA axis account for impaired cognitive function in this syndrome.

ACKNOWLEDGMENTS

The author would like to thank Christine Gall for helpful discussion and comments.

REFERENCES

[1] Elliott, E; Sapolsky, R. Corticosterone impairs hippocampal neuronal calcium regulation--possible mediating mechanisms. *Brain Res Bull.*, 1993 602, p. 84-90.

[2] Roy, M; Sapolsky, R. The exacerbation of hippocampal excitotoxicity by glucocorticoids is not mediated by apoptosis. *Neuroendocrinology*. 2003 77, p. 24-31.

[3] Sapolsky, RM; Puslinelli, W. Glucocorticoids potentiate ischemic injury to neurons: therapeutic implications. *Science,* 1985 229, p. 1397-1399.

[4] Sapolsky, RM; Krey, LC; McEwen, BS. Prolonged glucocorticoid exposure reduces hippocampal neuron number: implications for aging. *J. Neurosci.,* 1985 5, p. 1222-1227.

[5] Uno, H; Tarara, R; Else, J; Suleman, M; Sapolsky, R. Hippocampal damage associated with prolonged and fatal stress in primates. *J Neurosci,* 1989 9, p. 1705-1711.

[6] Watanabe, Y; Gould, E; Cameron, H; Daniels, D; McEwen, B. Phenytoin prevents stress- and corticosterone-induced atrophy of CA3 pyramidal neurons. *Hippocampus,* 1992 2, p. 431-435.

[7] Magarinos, A; McEwen, B; . Stress-induced atrophy of apical dendrites of hippocampal CA3c neurons: comparison of stressors. *Neuroscience,* 1995 69, p. 83-88.

[8] Radley, J; Rocher, A; Miller, M; Janssen, W; Liston, C; Hof, P; McEwen, B; Morrison, J. Repeated stress induces dendritic spine loss in the rat medial prefrontal cortex. *Cereb Cortex,* 2006 16, p. 313-320.

[9] Diamond, D; Bennett, M; Fleshner, M; Rose, G. Inverted-U relationship between the level of peripheral corticosterone and the magnitude of hippocampal primed burst potentiation. *Hippocampus,* 1992 2, p. 421-430.

[10] Sloviter, RS; Sollas, AL; Dean, E; Neubort, S. Adrenalectomy-induced granule cell degeneration in the rat hippocampal dentate gyrus: Characterization of an in vivo model of controlled neuronal death. *J. Comp. Neurol.,* 1993 330, p. 324-336.

[11] Sloviter, RS; Valiquette, G; Abrams, GM; Ronk, EC; Sollas, AL; Paul, LA; Neubort, S. Selective loss of hippocampal granule cells in the mature rat brain after adrenalectomy. *Science,* 1989 243, p. 535-538.

[12] Weiler, I; Spangler, C; Klintsova, A; Grossman, A; Kim, S; Bertaina-Anglade, V; Khaliq, H; de Vries, F; Lambers, F; Hatia, F; Base, C; Greenough, W. Fragile X mental retardation protein is necessary for neurotransmitter-activated protein translation at synapses. *Proc Natl Acad Sci U S A,* 2004 101, p. 17504-17509.

[13] Jin, P; Alisch, R; Warren, S. RNA and microRNAs in fragile X mental retardation. *Nat Cell Biol,* 2004 6, p. 1048-1053.

[14] Hessl, D; Dyer-Friedman, J; Glaser, B; Wisbeck, J; Barajas, R; Taylor, A; Reiss, A. The influence of environmental and genetic factors on behavior problems and autistic symptoms in boys and girls with fragile X syndrome. *Pediatrics,* 2001 108, p. E88.

[15] Lachiewicz, A; Dawson, D. Behavior problems of young girls with fragile X syndrome: factor scores on the Conners' Parent's Questionnaire. *Am J Med Genet,* 1994 51, p. 364-369.

[16] Bregman, J; Leckman, J; Ort, S. Fragile X syndrome: genetic predisposition to psychopathology. *J Autism Dev Disord,* 1988 18, p. 343-354.

[17] Cohen, I; Fisch, G; Sudhalter, V; Wolf-Schein, E; Hanson, D; Hagerman, R; Jenkins, E; Brown, W. Social gaze, social avoidance, and repetitive behavior in fragile X males: a controlled study. *Am J Ment Retard,* 1988 92, p. 436-446.

[18] Freund, L; Reiss, A; Abrams, M. Psychiatric disorders associated with fragile X in the young female. *Pediatrics,* 1993 91, p. 321-329.

[19] Mazzocco, M; Baumgardner, T; Freund, L; Reiss, A. Social functioning among girls with fragile X or Turner syndrome and their sisters. *J Autism Dev Disord,* 1998 28, p. 509-517.

[20] Miller, L; McIntosh, D; McGrath, J; Shyu, V; Lampe, M; Taylor, A; Tassone, F; Neitzel, K; Stackhouse, T; Hagerman, R. Electrodermal responses to sensory stimuli in individuals with fragile X syndrome: a preliminary report. *Am J Med Genet,* 1999 83, p. 268-279.

[21] Hessl, D; Glaser, B; Dyer-Friedman, J; Blasey, C; Hastie, T; Gunnar, M; Reiss, A. Cortisol and behavior in fragile X syndrome. *Psychoneuroendocrinology,* 2002 27, p. 855-872.

[22] Wisbeck, J; Huffman, L; Freund, L; Gunnar, M; Davis, E; Reiss, A. Cortisol and social stressors in children with fragile X: a pilot study. *J Dev Behav Pediatr,* 2000 21, p. 278-282.

[23] Bakker, C; Consortium, atD-BFX. FMR1 knockout mice:a model to study fragile X mental retardation. *Cell,* 1994 78, p. 23-33.

[24] Lauterborn, J. Stress induced changes in cortical and hypothalamic c-fos expression are altered in fragile X mutant mice. *Mol Brain Res,* 2004 131, p. 101-109.

[25] Lauterborn, JC; Rivera, S; Stinis, CT; Hayes, VY; Isackson, PJ; Gall, CM. Differential effects of protein synthesis inhibition on the activity dependent expression of BDNF mRNA transcripts: Evidence for immediate-early gene responses from specific promotors. *J Neurosci,* 1996 16, p. 7428-7436.

[26] Curran, T; Morgan, J. Fos: an immediate-early transcription factor in neurons. *J Neurobiol,* 1995 26, p. 403-412.

[27] Smith, MA; Makina, S; Kvetnansky, R; Post, RM. Stress and glucocorticoids affect the expression of brain-derived neurotrophin factor and neurotrophin-3 mRNAs in the hippocampus. *J. Neurosci.,* 1995 15, p. 1768-1777.

[28] Ueyama, T; Kawai, Y; Nemoto, K; Sekimoto, M; Tone, S; Senba, E. Immobilization stress reduced the expression of neurotrophins and their receptors in the rat brain. *Neurosci Res.,* 1997 28, p. 103-110.

[29] Cullinan, W; Herman, J; Battaglia, D; Akil, H; Watson, S. Pattern and time course of immediate early gene expression in rat brain following acute stress. *Neuroscience,* 1995 64, p. 477-505.

[30] Markham, J; Estrada, C; Greenough, W. Acute stress induces sex-specific changes in serum corticosterone in fragile X knock-out mice. *Soc Neurosci.,* 2005 115.15.

[31] Buckingham, J; Solito, E; John, C; Tierney, T; Taylor, A; Flower, R; Christian, H; Morris, J. Annexin 1: a paracrine/juxtacrine mediator of glucorticoid action in the neuroendocrine system. *Cell Biochem Funct.,* 2003 21, p. 217-221.

[32] Jessop, D. Review: central non-glucocorticoid inhibitors of the hypothalamo-pituitary-adrenal axis. J Endocrinol, 1999 160, p. 169-180.

[33] Voermans, P; Go, K; ter Horst, G; Ruiters, M; Solito, E; Parente, L. Expression of annexin and annexin-mRNA in rat brain under influence of steroid drugs. *Acta Neurochir Suppl,* 1997 70, p. 222-224.

[34] Sun, H-T; Cohen, S; Kaufmann, W. Annexin-1 is abnormally expressed in fragile X syndrome: two dimensional electrophoresis study in lymphocytes. *Am J Med Genet,* 2001 103, p. 81-90.

[35] Murphy, E; Spencer, R; Sipe, K; Herman, J. Decrements in nuclear glucocorticoid receptor (GR) protein levels and DNA binding in aged rat hippocampus. *Endocrinology,* 2002 143, p. 1362-1370.

[36] Sapolsky, R; Krey, L; McEwen, B. The adrenocortical axis in the aged rat: impaired sensitivity to both fast and delayed feedback inhibition. *Neurobiol Aging,* 1986 7, p. 331-335.

[37] Morano, M; Vazquez, D; Akil, H. The role of the hippocampal mineralocorticoid and glucocorticoid receptors in the hypothalamo-pituitary-adrenal axis of the aged Fisher rat. *Mol Cell Neurosci,* 1994 5, p. 400-412.

[38] Herman, J; Larson, B; Speert, D; Seaholtz, A. Hypothalamo-pituitary-adrenocortical dysregulation in aging F344/Brown-Norway F1 hybrid rats. *Neurobiol Aging,* 2001 22, p. 323-332.

[39] Pratt, W; Silverstein, A; Galigniana, M. A model for the cytoplasmic trafficking of signalling proteins involving the hsp90-binding immunophilins and p50cdc37. *Cell Signal,* 1999 11, p. 839-851.

[40] Reul, J; de Kloet, E. Two receptor systems for corticosterone in rat brain: microdistribution and differential occupation. *Endocrinology,* 1985 117, p. 2502-2511.

[41] Lauterborn, JC; Berschauer, R; Gall, CM. Cell-specific modulation of basal and seizure-induced neurotrophin expression by adrenalectomy. *Neuroscience,* 1995 68, p. 363-378.

[42] Lauterborn, JC; Poulsen, FR; Stinis, CT; Isackson, PJ; Gall, CM. Transcript-specific effects of adrenalectomy on seizure-induced BDNF expression in rat hippocampus. *Mol. Brain Res.,* 1998 55, p. 81-91.

[43] Imaki, T; Katsumata, H; Konishi, S-I; Kasagi, Y; Minami, S. Corticotropin-releasing Factor type-1 receptor mRNA is not induced in mouse hypothalamus by either stress or osmotic stimulation. *J. Neuroendocrin,* 2003 15, p. 916-924.

[44] Glass, DJ; Yancopoulos, GD. The neurotrophins and their receptors. *Trends Cell Biol.,* 1993 3, p. 262-268.

[45] Ebendal, T. Function and evolution in the NGF family and its receptors. *J. Neurosci. Res.,* 1992 32, p. 461-470.

[46] Murer, M; Yan, Q; Raisman-Vozari, R. Brain-derived neurotrophic factor in the control human brain, and in Alzheimer's disease and Parkinson's disease. *Prog Neurobiol,* 2001 p. 71-124.

[47] Huang, E; Reichardt, L. Neurotrophins: roles in neuronal development and function. *Annu Rev Neurosci,* 2001 24, p. 677-736.

[48] Horch, H; Krüttgen, A; Portbury, S; Katz, L. Destabilization of cortical dendrites and spines by BDNF. *Neuron,* 1999 23, p. 252-264.

[49] Murphy, D; Cole, N; Segal, M. Brain-derived neurotrophic factor mediates estradiol-induced dendritic spine formation in hippocampal neurons. *Proc. Natl. Acad. Sci. USA,* 1998 95, p. 11412-11417.

[50] Shimada, A; Mason, C; Marrison, M. TrkB signaling modulates spine density and morphology independent of dendrite structure in cultured neonatal Purkinje cells. *J. Neurosci.,* 1998 18, p. 8559-8570.

[51] Bramham, C; Messaoudi, E. BDNF function in adult synaptic plasticity: the synaptic consolidation hypothesis. *Prog Neurobiol,* 2005 76, p. 99-125.

[52] Ridder, S; Chourbaji, S; Hellweg, R; Urani, A; Zacher, C; Schmid, W; Zink, M; Hörtnagl, H; Flor, H; Henn, F; Schütz, G; Gass, P. Mice with genetically altered glucocorticoid receptor expressioon show altered sensitivity for stress-induced depressive reactions. *J. Neurosci.,* 2005 25, p. 6243-6250.

[53] Laggerbauer, B; Ostareck, D; Keidel, E; Ostareck-Lederer, A; Fischer, U. Evidence that fragile X mental retardation protein is a negative regulator of translation. *Hum Mol Genet,* 2001 10, p. 329-338.

[54] Li, Z; Zhang, Y; Ku, L; Wilkinson, K; Warren, S; Feng, Y. The fragile X mental retardation protein inhibits translation via interacting with mRNA. *Nucleic Acids Res,* 2001 29, p. 2276-2283.

[55] Zalfa, F; Giorgi, M; Primerano, B; Moro, A; Di Penta, A; Reis, S; Oostra, B; Bagni, C. The fragile X syndrome protein FMRP associates with BC1 RNA and regulates the translation of specific mRNAs at synapses. *Cell,* 2003 112, p. 317-327.

[56] Sung, Y; Dolzhanskaya, N; Nolin, S; Brown, T; Currie, J; Denman, R. The fragile X mental retardation protein FMRP binds elongation factor 1A mRNA and negatively regulates its translation in vivo. *J Biol Chem,* 2003 278, p. 15669-15678.

[57] Miyashiro, K; Beckel-Mitchener, A; Purk, T; Becker, K; Barret, T; Liu, L; Carbonetto, S; Weiler, I; Greenough, W; Eberwine, J. RNA cargoes associating with FMRP reveal deficits in cellular functioning in Fmr1 null mice. *Neuron,* 2003 37, p. 417-431.

[58] Johnson, L; Farb, C; Morrison, J; McEwen, B; LeDoux, J. Localization of glucocorticoid receptors at postsynaptic membranes in the lateral amygdala. Neuroscience, 2005 136, p. 289-299.

[59] Comery, T; Harris, J; Willems, P; Oostra, B; Irwin, S; Weiler, I; Greenough, W. Abnormal dendritic spines in fragile X knockout mice: maturation and pruning deficits. *Proc. Natl. Acad. Sci. USA,* 1997 94, p. 5401-5404.

[60] Weiler, I; Lin, L; Belt, B; Greenough, W. Some translational deficits in Fmr1-KO mice. *Soc. Neurosci.,* 2001 670.3.

[61] Li, J; Pelletier, MR; Perez Velazquez, JL; Carlen, PL. Reduced cortical synaptic plasticity and GluR1 expression associated with fragile X mental retardation protein deficiency. *Mol Cell Neurosci,* 2002 19, p. 138-151.

[62] Spencer, C; Alekseyenko, O; Serysheva, E; Yuva-Paylor, L; Paylor, R. Altered anxiety-related and social behaviors in the Fmr1 knockout mouse model of fragile X syndrome. *Genes Brain Behav,* 2005 4, p. 420-430.

[63] Peier, A; McIlwain, K; Kenneson, A; Warren, S; Paylor, R; Nelson, D. (Over)correction of FMR1 deficiency with YAC transgenics: behavioral and physical features. *Hum Mol Genet.,* 2000 9, p. 1145-1159.

[64] Qin, M; Kang, J; Smith, C. A null mutation for Fmr1 in female mice: effects on regional cerebral metabolic rate for glucose and relationship to behavior. *Neuroscience,* 2005 p. 999-1009.

[65] Mineur, YS; Sluyter, F; de Wit, S; Oostra, BA; Crusio, WE. Behavorial and neuroanatomical characterization of the Fmr1 knockout mouse. *Hippocampus,* 2002 12, p. 39-46.

[66] Zangrossi, HJ; Graeff, F. Behavioral validation of the elevated T-maze, a new animal model of anxiety. *Brain Res Bull.,* 1997 44, p. 1-5.

[67] Carvalho-Netto, E; Nunes-de-Souza, R. Use of the elevated T-maze to study anxiety in mice. *Behav Brain Res,* 2004 148, p. 119-32.

[68] Van Dam, D; D'Hooge, R; Hauben, E; Reyniers, E; Gantois, I; Bakker, C; Oostra, B; Kooy, R; De Deyn, P. Spatial learning, contextual fear conditioning and conditioned emotional responses in Fmr1 knockout mice. *Behav Brain Res,* 2000 117, p. 127-136.

[69] Trivedi, M; Coover, G. Lesions of the ventral hippocampus, but not dorsal hippocampus, impair conditioned fear expression and inhibitory avoidance on the elevaed T-maze. *Neurobiology of Learning and Memory,* 2004 81, p. 172-184.

[70] Kooy, R; Reyniers, E; Verhoye, M; Sijbers, J; Bakker, C; Oostra, B; Willems, P; Van Der Linden, A. Neuroanatomy of the fragile X knockout mouse brain studied using in vivo high resolution magnetic resonance imaging. *Eur J Hum Genet,* 1999 7, p. 526-532.

[71] Ivanco, T; Greenough, W. Altered mossy fiber distributions in adult Fmr1 (FVB) knockout mice. *Hippocampus,* 2002 p. 47-54.

[72] Diamond, D; Campbell, AP, CR; Vouimba, R-M. Preclinical research on stress, memory, and the brain in the development of pharmacotherapy for depression. *Eur Neuropsychopharm,* 2004 14, p. S491-S495.

[73] Brown, S; Rush, A; McEwen, B. Hippocmapal remodeling and damage by corticosteroids:implications for mood disorders. *Neuropsychopharm,* 1999 21, p. 474-484.

[74] McEwen, B. Plasticity of the hippocampus:adaptation to chronic stress and allostatic load. *Ann N Y Acad Sci,* 2001 933, p. 265-277.

[75] Koch, C; Zador, A. The function of dendritic spines: devices subserving biochemical rather than electrical compartmentalization. *J Neurosci,* 1993 p. 413-422.

[76] Koch, C; Zador, A; Brown, T. Dendritic spines: convergence of theory and experiment. *Science,* 1992 256, p. 973-974.

[77] Irwin, S; Patel, B; Idupulapati, M; Harris, J; Crisostomo, R; Larsen, B; Kooy, F; Willems, P; Cras, P; Kozlowski, P; Swain, R; Weiler, I; Greenough, W. Abnormal dendritic spine characteristics in the temporal and visual cortices of patients with fragile-X syndrome: a quantitative examination. *Am J Med Genet,* 2001 98, p. 161-167.

[78] Hinton, V; Brown, W; Wisniewski, K; Rudelli, R. Analysis of neocortex in three males with fragile X syndrome. *Am J Med Genet,* 1991 41, p. 289-294.

[79] Rudelli, R; Brown, W; Wisniewski, K; Jenkins, E; Laure-Kamionowska, M; Connell, F; Wisniewski, H. Adult fragile X Syndrome. Clinico-neuropathologic findings. *Acta Neuropathologica,* 1985 67, p. 289-295.

[80] Wisniewski, K; Segan, S; Miezejeski, C; Sersen, E; Rudelli, R. The Fra(X) syndrome: neurological, electrophysiological, and neuropathological abnormalities. *Am J Med Genet,* 1991 38, p. 476-480.

[81] Woolley, C; Gould, E; McEwen, B. Exposure to excess glucocorticoids alters dendritic morphology of adult hippocampal pyramidal neurons. *Brain Res,* 1990 531, p. 225-231.

[82] Wellman, C. Dendritic reorganization in pyramidal neurons in medial prefrontal cortex after chronic corticosterone administration. *J Neurobiol,* 2001 49, p. 245-253.

[83] Shors, T; Chua, C; Falduto, J. Sex differences and opposite effects of stress on dendritic spine density in the male versus female hippocampus. *J. Neurosci.,* 2001 21, p. 6292-6297.

[84] Sunanda Rao, M; Raju, T. Effect of chronic restraint stress on dendritic spines and excrescences of hippocampal CA3 pyramidal neurons--a quantitative study. *Brain Res,* 1995 694, p. 312-317.

[85] Seib, L; Wellman, C. Daily injections alter spine density in rat medial prefrontal cortex. *Neurosci. Letts.,* 2003 337, p. 29-32.

In: New Research on Fragile X Syndrome
Editor: Jeffrey J. Upner, pp. 97-115

ISBN: 1-60021-281-6

Chapter 6

GENETIC COUNSELLING IN FRAGILE X SYNDROME: HOW NEW DISCOVERIES ABOUT THE SYNDROME HAVE COMPLICATED IT

Laia Rodriguez-Revenga and Montserrat Milà
Hospital Clínic, Barcelona, Spain.

ABSTRACT

Fragile X syndrome is the most common form of inherited mental retardation. Molecular basis of the syndrome is a CGG expansion in 5' untranslated region of the *FMR1* gene. It is inherited as X linked dominant trait, with a reduced penetrance (80% for males and 30% for females). The general mechanism by which the unstable region is transmitted to the next generation has been established: full mutation expansion from premutated alleles is only acquired via maternal meiosis, with an increased risk for the larger premutation sized alleles. On the other hand, paternal transmission always remains in the premutated range, from both premutated and full-mutated fathers. Genetic counseling and prenatal diagnosis for this syndrome have substantiality improved using molecular testing. Nowadays and thanks to the new advances in assisted reproductive techniques such as donor oocytes, preimplantational diagnosis or prenatal diagnosis, the reproductive options for carrier females have considerably increased. The knowledge of the precise CGG repeat number in premutated carriers, allows us to perform an accurate genetic counseling.

Moreover, new advances in fragile X syndrome have resulted in the description of pathologies associated to premutation carriers. Between 12-20 % of premutated women present premature ovarian failure (menopause before 40 years), while only 1% of general population presents it. On the other hand, one third of normal transmitting males and an unknown percentage of premutated women, present a new syndrome of tremor ataxia associated to fragile X premutation (FXTAS) after 60 years. These findings have modified genetic counseling for fragile X syndrome, as they should be taken into account.

In this section, we analyze the genetic counseling given to 300 fragile X families through more than 10 years of experience. Who, when and where have to receive the information and how it has to transmitted to the rest of the family. We report the reproductive options chosen by families on basis of our 150 prenatal diagnoses experience. Finally, we also explain the impact received by the families when the information about premature ovarian failure and FXTAS syndrome is given.

In summary, this section shows our experience in genetic counseling, prenatal diagnosis and associated pathologies for 300 fragile X families.

FRAGILE X SYNDROME

Fragile X syndrome (FXS) is the leading form of hereditary mental retardation, but the condition is still underdiagnosed in young children, since the phenotype at this age is very subtle.

In 1943 Martin & Bell described a form of familial mental retardation, which affected males, and it was associated to dysmorphic traits and machroorchidism [1]. Twenty-six years later, Lubs associated a cytogenetic marker that consisted in an X chromosome, which was broken at the end of its long arm [2]. Since then the syndrome was called fragile X. Finally, in 1991 three groups working separately cloned a gene named *FMR1* (Fragile Mental Retardation type 1), which is responsible for the syndrome [3-5].

Nowadays, FXS is considered the commonest form of inherited mental retardation. The real incidence of the syndrome is not known, but epidemiological studies indicate that it is responsible for mental retardation in 1 in 4,000-6,000 males and in 1 in 7,000-10,000 females of European descendent. In a recent study performed in Catalonia it gave an incidence of 1:2466 male and 1:8333 females [6]. This difference between males and females is due to reduced penetrance: 80% of males and 30% of females and the phenomena of X inactivation in females. It is also important to highlight the high incidence of premutation carriers, 1 in 1233 males and 1 in 411 females [6].

The knowledge of the *FMR1* gene and the molecular defect causing FXS help us to understand the genetic aspects of the syndrome. Although the syndrome is inherited as a dominant X-linked trait, it is characterized by an unusual mode of transmission. One frequently finds in affected families clinically normal transmitting males (NTM), whose daughters – also clinically normal - have a high risk of having affected children [7]. The risk of developing the disease increases in successive generations of the same family through maternal transmission [8].

The *FMR1* gene is located in the long arm of the X chromosome at Xq27.3, it expands 17 exons, and 40 kb of genomic DNA. It transcribes an mRNA of 3.9 kb, and the translated protein is called FMRP. The lack of this protein is the cause of the FXS. The molecular basis of the syndrome in the majority of cases is a dynamic mutation, an expansion and hypermethylation of an unstable CGG trinucleotide repeat located in the first exon of the *FMR1* gene [3-5]. In a normal situation the number of CGG repeats is polymorphic with alleles between 6 to about 52. In this situation the CpG island, which is located in the promoter of the gene acting as a switch depending on its methylation status, is non-methylated. Thus the gene is active, *FMR1* is transcribed and translated. Repeats at this size remain stable upon transmission. A second class of alleles that overlaps with the upper range of the wild-type is those with ~40-55 CGGs. This range, known as the gray zone, never expands to full mutation but it is transmitted slightly unstable to subsequent generations with the possibility of creating a premutated allele. In the premutation, alleles range from 55 to about 200 CGG repeats. At this situation, the *FMR1* gene is also transcribed as the CpG island is non-methylated. Therefore, premutated carriers have normal or lightly reduced synthesis of FMRP and they are asymptomatic for FXS. However, they have risk of having affected descendence since the number of CGG is unstable, and in each cellular division it can increase, and can be transmitted with a higher number of repeats to the next generation. Finally, when the CGG number is over 200 repeats (full mutation), the CpG island is methylated and this methylation switches off the gene, blocking transcription, and no protein is translated. These individuals are always affected if they are males and in about 30% of females. Moreover, about 15% of patients show a mosaic pattern, referring to a mixture of premutation and full mutation, which is caused by the somatic instability of the full mutation in the early embryogenesis that can lead to retraction of the expanded repeat. Another type of mosaicism can be observed in large expansions with incompletely methylation. These two types of mosaic mutations allow the expression of some FMRP and in some cases they have been associated with milder mental retardation in males [8].

Although it is well known that the lack of FMRP is the responsible for all clinical phenotype in the FXS patients, the exact function of this protein is still undefined. FMRP expression is widespread and abundant in neurons as well as spermatogonia [9-11]. Its subcellular distribution is largely cytoplasmatic, highly concentrated in association with polyribosomes and rough endoplasmic reticulum, including those present in dendrites [12]. FMRP contains several RNA-bindings motifs, including two KH domains and an RGG box [13, 14]. In addition, a nuclear localization signal, a nuclear export signal, two coiled coils and a G-quartet binding structure have been identified. On the basis of its properties and its localization, it is suggested that FMRP plays a role in the regulation of transport, stability or translation of some mRNAs from nucleus to cytoplasm, in synapse maturation during development, and for synaptic activity in the adult. Interestingly, recent evidence shows that mRNA transport in dendrites plays an important role in neuronal process, including those essential for memory consolidation and learning process [12, 15, 16]. In fact, examination of autopsy material from FXS patients and brains of *Fmr1* knock-out mice revealed immature synaptic connections [17].

In vitro binding studies revealed that FMRP binds to a variety of RNA molecules, including a subset of brain and its own mRNAs. However, the RNA targets to which FMRP binds *in vivo* remains elusive. The knowledge of these interacting proteins is needed in order to understand the real function of FMRP.

THE FRAGILE X PREMUTATION

Carriers of premutation expansions (55 to 200 CGG repeats) of *FMR1* gene generally have been considered to have a normal cognitive functioning [18-21]. Even though, and contrary to expectation, there are reported several subgroups of male and female premutation carriers displaying features consistent with the typical clinical spectrum of FXS. For instance, Riddle and co-workers [20] and Hagerman [22] described a group of premutation carriers with prominent ears and joint laxity; or with learning disabilities, attention deficit/hyperactivity disorder (ADHD), or difficulty with math [20, 23]. In all cases, such findings are consistent with a lowered FMRP levels observed in the peripheral blood leucocytes, particularly in those with larger CGG expansions (>100 repeats) [24].

There are, however, two forms of clinical involvement among carriers of premutation alleles that are not consistent with the clinical spectrum of the FXS. These two disorders are the premature ovarian failure (POF) and the fragile X-associated tremor/ataxia syndrome (FXTAS). Both disorders have in common that the resulted phenotype does not share anything with the FXS clinical spectrum. In other word, these two newly disorders are independent with the FXS, and therefore any affected patient (with a full mutated allele) will never develop a POF or a FXTAS syndrome.

Premature Ovarian Failure

The average age for menopause in Western populations of women is approximately 51 years. Premature ovarian failure (POF) or premature menopause, which affects approximately 1% of women from the general population, refers to development of amenorrhea due to cessation of ovarian function before the age of 40 years. The diagnosis is based on elevated FSH levels in menopausal range (usually above 40IU/I) detected on at least two occasions a few weeks apart [25, 26].

Women with POF suffer from anovulation and hypoestrogenism and present with primary or secondary amenorrhoea, infertility, sex steroid and elevated gonadotrophins. Early loss of ovarian function results in significant psychosocial squeals and major implications, since it has been reported that nearly 2-fold age-specific increase in mortality [26, 27].

A wide spectrum of pathogenic mechanisms may lead to the development of POF including chromosomal, genetic, autoimmune, metabolic (galactosaemia), infectious (mumps) and iatrogenic (anticancer treatments) causes. Although most cases of POF are idiopathic, observation of familial cases with POF indicates a genetic role in its pathogenesis [26, 28].

Historically, at the FXS Conference in Denver in 1987, one of the first meetings that combined families with the FXS and scientists, a spontaneous discussion occurred among heterozygous carriers concerning the early onset of their menopause [29]. Lately, by the early 1990s, POF was first noted as an unexpected phenotype among premutation carriers of the *FMR1* gene and, subsequent studies firmly associated the FRAXA premutated alleles with the presence of POF. This evidence has been demonstrated in both, by the analysis of women carrying the premutation and by screening women affected with POF [30-35].

Several studies examining premature menopause in women from FXS families have been performed. The results from a large international collaborative study that examined premature menopause in 760 women from FXS families ranged POF among premutation carriers from 16% (inclusion of all women ≥18 years) to 24% (inclusion of all women ≥40 years). On the other hand, none of the 238 full mutated and one (0.4%) of the 237 controls experienced POF, a frequency similar to that of the general population [36]. If the cessation of menses before age 40 is seen in ~20% of premutated females, for those who have not yet gone through menopause, elevated FSH (indicating an ovarian dysfunction) is seen in ~25% of female carriers aged <40 years [37].

The incidence of FRAXA premutation has also been shown to vary among women with POF depending on the proportion of sporadic and familial cases. Thirteen per cent of pedigrees with the familial POF and 3% of women with the sporadic form of POF have been found to carry premutated alleles compared with an expected prevalence 1:590 [26, 32].

It is not clear if there is a variable expressivity or a reduced penetrance among premutation carriers that would explain why not all premutated women develop a POF. Hundscheid and co-workers [38] reported evidence for a paternal-parent-of-origin effect on POF in female premutated carriers, which might support the reduced penetrance model. Carriers who received the premutation from their fathers were at a higher risk of POF (28%) than those who received the premutated allele from their mothers (4%). However, subsequent studies by other authors find both maternal and paternal inheritance [35, 39-41]. Another interesting point regarding premutation and POF, is that there is no evidence that correlates the menopause age and the repeat size or the ratio of X chromosome inactivation [42].

The underlying mechanism for explaining the association between POF and FRAXA premutated alleles is still not clear. Hypotheses related to the *FMR1* transcript and the protein level has been proposed. It has been described that the *FMR1* gene is expressed in several areas that may affect stages of follicular development. In a previous work performed with ovary samples from full mutated female fetus, our group reported FMRP expression in all germ cells while the paragranulosa and interstitial cells were FMRP-negative. The Mullerian epithelium of the fetal Fallopian tube was continuously positive in the control case, whereas the full mutated sample showed discontinuous pattern [43]. In a murine study, Bachner and co-worker [44] detected enhanced levels of *fmr1* expression in the fetal ovary while in the mature ovary no specific expression was found. As proliferation of oogonia takes place in the fetal ovary, it was suggested that *FMR1* serves a special function during germ cell proliferation [26]. Interestingly, it has been described that premutated carriers have increased levels of *FMR1* mRNA and decreased FMRP levels. Although it is not well understood, these subtle changes in FMRP expression may affect ovarian function. For all these reason, further investigations are needed to understand the molecular pathways underlying POF among premutated carriers.

On the basis of all these findings, the genetic counseling for premutated women has to include a fertility advice. On average, menopause for all premutated carriers comes earlier, and therefore delaying the reproduction may not be a good option. Moreover, women with premutation may not be good candidates for fertility treatment [41]. All this data should be taken into account when a premutated carrier is seeking genetic advice.

The Fragile X-Associated Tremor/Ataxia Syndrome

This second form of clinical involvement among FRAXA premutated carriers has been described more recently [45]. It was firstly identified among older male carriers of premutation alleles, and includes progressive action tremor and ataxia with associated radiological findings [24]. However, to date FXTAS has also been described among premutated women although it has been suggested that it occurs less frequently and that the phenotype is milder with older age at onset [45, 46]. An explanation for this difference, is the presence of a second normal allele and a random X-inactivation of the premutated one; however, there may be additional sex-specific effects that reduce penetrance among females [24].

Clinical symptoms of FXTAS syndrome appear in patients in their 50s or later, and they usually begin with an action tremor. After that, different findings including ataxia (balance problems with frequent falling), and more variably loss of sensation in the distal lower extremities and autonomic dysfunction (e.g., impotence, hypertension, and loss of bowel and bladder function) gradually progress. Cognitive deficits are also observed and include memory problems and executive function deficits with a gradual progression to dementia in some individuals. Several of the patients have parkinsonian symptoms, including an intermittent resting tremor, masked facies, and increased tone. The patients also appear to experience psychological symptoms, including anxiety, reclusive behavior, and irritability or mood lability, all of which are likely to be related to the cognitive changes or to changes in the limbic area of the brain [24]. Magnetic resonance imaging (MRI) of patients affected with FXTAS demonstrates mild to moderate brain atrophy, which includes the cerebellum, with dilated ventricles and scattered perventricular white matter disease. In addition, hyperintensities of the middle cerebellar peduncles (MCP) on T2 is characteristic of FXTAS patients [24, 47]. In a work performed by Jacquemont et al. [48] there are described three diagnostic categories in the diagnosis of FXTAS, each one reflecting combination of symptom types (Table 1). The diagnostic categories are: i) "Definite" indicates the presence of one major radiological sign plus one major clinical symptom; ii) "Probable" indicates the presence of either one major radiology sign plus one minor clinical symptom or two major clinical symptoms; and iii) "Possible" indicates the presence of one minor radiological sign plus one major clinical symptom.

It has been estimated that at least one-third of all male carriers will develop a FXTAS syndrome, although there is significant variability in the progression of neurological dysfunction [24, 46, 49]. In our FXS families the estimated prevalence of FXTAS is 20% (data unpublished).

Table 1. Clinical criteria for FXTAS.

Examination and Degree	Observation
Radiological:	
Major	MRI white matter lesions in MCPs and or brain stem
Minor	MRI white matter lesions in cerebral white matter
Minor	Moderate-to-severe generalized atrophy
Clinical:	
Major	Intention tremor
Major	Gait ataxia
Minor	Parkinsonism
Minor	Moderate-to-severe short-term memory deficiency
Minor	Executive function deficit

Inclusion criterion: CGG repeat number between 55 and 200.
Note. Data described by Jacquemont *et al.* [48].

The description and characterization of FXTAS syndrome is of great interest to the population, because the prevalence of *FMR1* premutation in the general population is relatively high. In fact, it as been estimated that 1:1233 males and 1:411 females are a FRAXA premutated carrier [6]. Several studies have been performed in order to determine the real role of FXTAS in undiagnosed adult patients with movement disorders. The results obtained in European populations ranges from 0% to 4% [50-53]. Our studies show an estimated FXTAS prevalence of 2% among patients presenting with ataxia of unknown etiology (data unpublished). Although large studies are necessary to better define FXTAS prevalence in this kind of population, on the basis of premutation male frequency in general population, the prevalence of FXTAS has been estimated in ~1/3,000 males aged over 50 years of age (~1/10,000 males of all ages) [24].

FXTAS is an allelic disorder to the FXS, and therefore should be considered as a distinct neurodegenerative disorder. In fact, the molecular mechanism leading to FXTAS is distinct from the *FMR1* silencing mechanism and/or a deficit in FMRP operating in FXS. In premutated patients the *FMR1* gene is rarely silenced and FMRP levels are generally normal or only slightly lowered [24]. The only known molecular abnormality among premutation carriers is the presence of elevated levels of *FMR1* mRNA. The increased transcriptional activity of the *FMR1* gene seems to be positively correlated with the size of the CGG repeat. That is, CGG repeats in the upper range (100-200 CGG) result in average 5-fold elevation, whereas CGGs in the lower range (50-100 CGG) result in an average 2-fold elevation [54-57]. The presence of these elevated levels of abnormal (expanded CGG repeat) *FMR1* mRNA led to propose an RNA "toxic gain-of-function" model for FXTAS, in which the mRNA itself is causative of the neurological disorder [24, 45, 48, 58]. The same mechanism has been proposed for myotonic dystrophy (DM1 and DM2), in which either the expanded repeat tract of CUG in DM1 or CCUG in DM2 sequestered CUG-binding proteins that disrupts mRNA processing of other genes or transport of other mRNAs [59]. In fact, this model has been demonstrated for DM1 by placing an expanded CTG tract in 3' UTR region of the *DMPK* mRNA in a transgenic mouse [60]. A part from this finding, FXTAS and myotonic dystrophy have another important similarity that will support the RNA gain-of-function mechanism for FXTAS syndrome. Both disorders show nuclear inclusions produced as a result of the binding proteins sequestered by the respectively mRNA, with a cytotoxic effect that lead to cell death. In a study performed by Greco and co-worker [58] eosinophilic intranucelear inclusions in neurons and astrocytes throughout the cortex and in deep cerebellar nuclei of FXTAS post-mortem samples were reported. Furthermore, in a subsequent study, there is described a highly significant association between CGG length and both the number of inclusions and the age of death, which correlates with the progressive character of the disease [61].

In order to test this RNA gain-of-function hypothesis for FXTAS, a mouse model has been generated in which the endogenous mouse CGG repeat was replaced by a human CGG tract carrying 98 CGGs [62]. Further studies of the brain of these expanded-repeat mice (at 20-72 weeks) evidenced elevated *Fmr1* mRNA levels and ubiquitin-positive intranuclear inclusions [63]. An increase was also observed in both the number and the size of the inclusions in specific brain region during the course of life [57]. The presence of inclusions in this mouse, that has normal levels of FMRP, provides evidences against a protein-deficiency model for FXTAS, and supports a direct role of the *Fmr1* gene, by either CGG expansion *per se* or by elevated *Fmr1* mRNA levels, in the pathology as the gain-of-function mechanism hypothesized.

To conclude, as the prevalence of premutated alleles is relatively high in general population, FXTAS may represent one of the more common monogenic causes of tremor, ataxia, and dementia. For this reason, it is probably that many carriers with FXTAS are being seen by a clinical specialist without awareness of the underlying genetic basis for the symptoms. The early diagnosis of those patients not only benefits themselves but also the rest of the family that should be advised for the FXS.

Genetic Counselling

In the FXS and like an X-linked disorder the chances that the son or daughter will inherit the mutated chromosome from the mother are 50%, while all the daughters and none of the sons of carrier fathers receive the mutation. However, if the child will be affected or unaffected it depends on the dynamic mutation process, if the mutation remains in premutation range or progresses to the full mutation. The interested individual should receive FXS genetic counseling in person and by an expert geneticist. It is important to highlight that in some situations, the diagnosis of FXS in a family reveals premutation carrier status of all relatives linked to the affected individual, thereby revealing an individual's risk for a late age disorder, POF and/or FXTAS. At this situation, it might arise some ethical problems such as the choice of "knowing/not knowing" this carrier status or when the family members of the FXS diagnosed patient refuse to explain it to at-risk relatives.

Females with Full Mutations

If the mother carriers a full mutation, 50% of her offspring will inherit the mutation. Although rare, there are reports of women with full mutations having offspring with contractions or premutation-sized alleles [64, 65]. If it is a son, who inherits the mutation, he will be affected of FXS, while if it is a daughter no prediction of the clinical and severity of the phenotype can be made. Up to 50% of females with full mutations have some of the physical features associated with FXS. Intellectual impairment is often milder in females than in affected males, and cognitive functioning can range from normal intelligence to learning disabilities to mental retardation [66]. It has been estimated that 53-71% of females with full mutations have IQs in the borderline or mentally retarded range [67-69]. Our data, however, shows that mental retardation needing special schools affects only 30% of them [6]. For this reason, nowadays and despite all the advances in the diagnostic methodologies, the genetic counseling for those females who inherit a full mutation remains challenging.

Females with Premutation

When the mother is a premutated carrier, the chance that the premutation will be converted to full mutation depends on several aspects such as the number of CGG repeats, the gender of the carrier, and the presence of AGG interruptions within the repeat. The process by which CGG repeat expansion occurs may be related to difficulties in DNA replication of the repeat. Premutations may undergo expansions during oogenesis in carriers [70], and during postzygotic mitosis in children who inherit the premutation from their mothers [66, 71, 72].

Several studies have been performed in order to estimate the risk of expansion for different CGG repeat ranges (Table 2). It is important to note that no expansion to full mutation in one generation has been reported with alleles containing fewer than 59 CGG repeats [73, 74]. On the basis of this data, although we always advice prenatal diagnosis to all female carriers, we inform them that those women carrying alleles between 53 to 59 repeats have a very low probability of expansion to full mutation; nevertheless, they always choose to perform prenatal diagnosis.

Table 2. Risk of CGG expansion for fragile X premutated female carriers.

Maternal repeat size	Nolin *et al.*, [73]	Rife *et al.*, [74]
55-59	4 (1/27)	16.6 (2/12)
60-69	5 (6/113)	21 (7/33)
70-79	31 (28/90)	41.1 (21/51)
80-89	58 (81/140)	59.2 (45/76)
90-99	80 (89/111)	96.1 (25/26)
100-200	98 (194/197)	100 (106/106)

A part from the advice of the chance of having an affected child, genetic counselor has to be aware that females with premutations are at increased risk (around 20%) for POF. Those premutated females diagnosed before the description of POF findings, should be informed about it. To summarize, genetic counseling to premutated females should include a fertility and reproductive mention (see the section on Reproductive Options). Although, the involvement of FXTAS in women has been far less found when comparing with men, it must be considered for those presenting neurological symptoms over 60 years of age.

Males with Full Mutation

Although affected males have difficulty in social integration, there exist high functioning males (HFM). These are males carrying full mutations but with the CpG island and the CGG repeats unmethylated, resulting in a normal *FMR1* gene function. Genetic counseling for these males is good, since full mutations are only transmitted via the female germline. In fact, it has been observed that the sperm of full-mutation males have only premutations [74, 75], and therefore a male with full mutation will always have premutated daughters.

Males with Premutation

The risk of expansion to the full mutation from carrier men to their daughters is almost zero. Males with premutation pass on the premutation to all of their daughters, usually with only small expansions or contractions. All of his daughters will be normal carriers but with a high risk of having affected children in the next generation. Moreover, the recognition of FXTAS as a new neurological disorder associated with older premutated male carriers has to be considered in the genetic counseling. Preliminary studies suggested that FXTAS occurs in ~20-30% of premutated male carriers [57]. At this time, we are contacting with all our FXS families in order to inform them about FXTAS syndrome.

Reproductive Options

Individuals at risk for passing on FXS mutations to their offspring have a variety of pre- and postconception options available [66]. When offering the reproductive options to a couple at risk if the premutated carrier is the female a fertility issue has to be considered since the risk of POF has significant reproductive implications. First of all, the onset of POF is difficult to predict. Secondly, the subtle endocrine perturbations (elevated follicle-stimulating hormone levels) observed in these women decreases the efficiency of ovarian stimulation required for preimplantational diagnosis. Finally, there is evidence that premutation carriers may have hormonal changes suggestive of early ovarian aging despite regular menstrual cycles [66, 76].

Offspring Renounce and Adoption

These two options are drastic, and therefore they are nowadays seldom observed as the first choice option by couples at risk that wish to prevent the birth of an affected child. The options of remaining childless and the adoption are usually rejected for the couples.

Donor Germline Cell

Pregnancy can be achieved using donor eggs or sperm for female and male mutation carriers, respectively. Nowadays, it is a very good option for premutated females who have been involved in previous prenatal diagnosis and termination of pregnancies. Given the relatively high prevalence of FXS in the general population, potential gamete donors should be tested for this syndrome. Although the efficiency and reproducibility of sex selection techniques are controversial, males at risk for passing FXS premutations to their offspring may wish to consider sperm-sorting techniques for sex selection males [66]. Based on our experience, around 5% of FXS families chose for a germline cell donation.

Prenatal Diagnosis

The prenatal diagnosis is the first choice selected by FXS families. We have estimated that around 95% of our families have chosen this reproductive option, since it avoids the absence of a genetic parental relationship that results after an adoption or after an germline cell donation. Fetal testing performed on amniotic fluid cells or chorionic villi is available when parent is a known premutation or full mutation carrier. The CGG number in the fetus can be determined with precision, but it is advisable to perform both PCR and Souther blot analysis. Although prenatal diagnosis can be done by using DNA sample obtained by either amniocentesis or chorionic villus sampling, we prefer the second one since more DNA quantity is obtained and because the prenatal diagnosis can be performed earlier. This fact reduces the anxiety of the couple and it allows interrupting the pregnancy earlier, which reduces the risk of complications. However, the methylation pattern observed in the chorionic villi sample can be incomplete, and therefore may not be the one found in the liveborn [77]. In order to solve this problem, chorionic villi sample should be obtained at 13 weeks of gestation, when methylation patterns are already established.

In our Genetics Service we have given genetic counseling to 300 FXS families through more than 10 years of experience. During this period of time, we have performed 150 prenatal diagnoses from premutated carriers or full mutated females. Among them, 37 diagnoses were performed to females with family history of mental retardation of unknown etiology. As the status of the mother was unknown, those prenatal diagnoses were performed while studying the pregnant. All of them were normal for the *FMR1* gene expansion. The 113 remaining prenatal diagnoses were done within known FXS families. As summarized in Table 3, 47 cases (21 males and 26 females) inherited the normal *FMR1* allele from the carrier progenitor. In 11 cases (5 males and 6 females) the embryo inherited the paternal premutated allele, but in all cases this allele remained at the premutation range. Finally, 55 cases (25 males and 30 females) showed a full mutated allele. In all affected male embryos the couple decided to interrupt the pregnancy, however in two female cases, the pregnancy was carried on. Usually when the couple has an affected son or daughter they decide to terminate the pregnancy, but in those cases in which the mother presents a full mutated allele, they decide to continue with it. This is the situation found in the two affected female prenatal diagnoses in which the couple decided to carry on with the pregnancy.

Table 3. One hundred thirteen prenatal diagnoses performed among FXS families.

STATUS / SEX	FULL MUTATED	PREMUTATED	NORMAL RANGE	TOTAL
MALES	25	5	21	51
FEMALES	30	6	26	62
TOTAL	55	11	47	113

We always perform cytogenetic studies in all prenatal diagnoses, not only because it allows us to know the sex of the fetus, but also because the coexistence with other genetic pathologies is a possibility. In fact, we have described a prenatal diagnosis in which the molecular testing for FXS disclosed a normal male for this pathology but the cytogenetic analysis evidenced a structural abnormality affecting the long arm of chromosome 18 [78].

A comprehensive genetic counseling for FXS couples in whom a prenatal diagnosis is going to be carried out should be given. This might include the benefits and limitations of an invasive procedure and the implications of prenatal results, including the range of possible outcomes for a female fetus with a full mutation [66]. The variable phenotype among males and females with premutation should also be emphasized, as well as the possibility of a methylation mosaicism when an NTM is detected.

Finally, we have to mention that prenatal diagnosis and the further termination of pregnancy if the fetus is affected produce emotional and psychological problems to couples. For this reason the development of new techniques such as the preimplantation genetic diagnosis are being highly demanded.

Preimplantational Diagnosis

Preimplantation genetic diagnosis (PGD) is increasing among FXS families because the probability of finding a fetus with a full mutation is 50%. It is more demanded when the couple has passed through previous prenatal diagnosis and interruption of affected pregnancies. PGD can be considered as an early form of prenatal diagnosis carried out before embryonic implantation. Hence, the physical and psychological trauma associated with a possible termination of pregnancy is avoided [79]. PGD for FXS is possible but there are some important aspects that have to be considered. The first one is that the diagnosis is based on the presence of the normal allele. To be able to perform a FXS PGD the number of triplet repeats on the healthy *FMR1* allele of the couple at risk have to be informative and different from each other, which only occurs in 63% of couples [80]. This way, non-carrier female embryos and unaffected male embryos can safely be selected for transfer. Premutation carrier embryos will only be identified if the numbers of repeats do not exceed 75. Larger premutations or full mutations will not be seen due to PCR technique limitations, and therefore this means that a number of embryos carrying a premutation will be lost for transfer [79]. Linked polymorphic markers can also be used in order to help those non-informative couples. As mentioned before, the second aspect that has to be considered on PGD is that those women carrying a premutation have a reduced ovarian response to stimulation protocols resulting in few embryos for biopsy and diagnosis. In fact, in a study performed by Platteau and co-workers [79], they found that the volume of embryos retrieved was reduced, and that only 55% of fertilized oocytes reached the stage of embryo biopsy, thus leading to a significant reduced opportunity for pregnancy.

In some countries the PGD has two additional limitations: *i*) few laboratories are capable of offering the technique, and *ii*) the amount of time that couples have to wait for an assisted reproduction technique. These two aspects together with the technical limitations of the diagnostic procedure and the poor ovarian response explain why none of our FXS families have succeeded with PGD.

POPULATION SCREENING

The implementation of FXS screening in the general population is an issue on debate due to the high frequency of the syndrome and premutated carriers. Several population screenings have been proposed, depending on the individuals that are screened (newborn, women of reproductive age, pregnant women, and children with developmental delay). The genetic counseling that accompanies each population screening is highly important and might be different depending on how and when an individual is identified. Moreover, we have to be aware that the implications of a positive test will not be the same for each tested population.

During the 2002, we performed a pilot study among a group of 200 newborn males in which we detected one full mutated [81]. The advantages and disadvantages of this kind of newborn screenings are still under consideration. Surely, the major benefit of them is the early detection of FXS patients, which is important in order to provide early intervention and supportive care and to inform parents for further family planning. Screening of the general population for identification of carrier females and offering them prenatal diagnosis for prevention of the birth of a FXS child, have also been tried by other countries [82]. In a recent study performed by Anido and co-workers [83] it is described that the timing of the carrier testing with respect to a woman's life stages dictates whether the information on carrier status will be seen as beneficial or detrimental. They also found evidence that non-carrier women from the general population would be wholly unprepared for positive carrier result [83].

A population carrier screening would also detect women with intermediate-sized alleles. In such cases, genetic counseling should emphasized that these kind of individuals are not generally considered to be at risk for clinical manifestations of either pre- or full mutations [66].

As exposed, there are still a lot of controversial data that argue against population screenings, especially among pregnant and nonpregnant women.

Conclusion

The knowledge of the *FMR1* gene and the molecular defect causing FXS has highly improved FXS diagnosis and genetic counseling. However, there are some particular aspects of the syndrome, including the unusual mode of the disease transmission, and the variable phenotype that full mutated females can present that difficult it. Moreover, the description of associated pathologies to premutation carriers, POF and FXTAS, has also modified genetic counseling for FXS. Nowadays, these two disorders and their consequences have to be taken into account by genetic counselors. Finally, care must be taken before the implementation of any population screening. For this reason, nowadays FXS testing is only recommended for: *i*) individuals of either sex with mental retardation, developmental delay, or autism especially when associated with other physical and behavioral characteristics of FXS, a family history of FXS, or a relative with undiagnosed mental retardation; *ii*) individuals with a family history of FXS or a family history of undiagnosed mental retardation who are seeking reproductive counseling; *iii*) women with reproductive or fertility problems, especially if there is a family history of POF, FXS, or undiagnosed mental retardation; and *vi*) individuals with late onset ataxia of unknown etiology, particularly when there is a family history of movement disorders, FXS, or undiagnosed mental retardation.

Acknowledgements

This work has received financial support from GIRMOGEN (V2003REDG 03-98), RECGEN (V2003REDC 07), PI050776, PI050159, financed by "Instituto Carlos III", and SAF2004-03083 financed by the "Ministerio de Educación y Ciencia, Spain.

References

[1] Martin, JB & Bell, J. A pedigree of mental defect showing sex-linkage. *J Neurol Psychiatr* 1943;6:154-157.

[2] Lubs, HA. A marker X chromosome. *Am J Hum Genet* 1969;21:231-244.

[3] Oberle, I; Rousseau, F; Heitz, D; Kretz, C; Devys, D; Hanauer, A; Boue, J; Bertheas, MF; Mandel, JL. Instability of a 550-base pair DNA segment and abnormal methylation in fragile X syndrome. *Science* 1991;252:1097-1102.

[4] Verkerk, AJ; Pieretti, M; Sutcliffe, JS; Fu, YH; Kuhl, DP; Pizzuti, A; Reiner, O; Richards, S; Victoria, MF; Zhang, FP; *et al.* Identification of a gene (FMR-1) containing a CGG repeat coincident with a breakpoint cluster region exhibiting length variation in fragile X syndrome. *Cell* 1991;65:905-914.

[5] Yu, S; Pritchard, M; Kremer, E; Lynch, M; Nancarrow, J; Baker, E; Holman, K; Mulley, JC; Warren, ST; Schlessinger, D; *et al.* Fragile X genotype characterized by an unstable region of DNA. *Science* 1991;252:1179-1181.

[6] Rife, M; Badenas, C; Mallolas, J; Jimenez, L; Cervera, R; Maya, A; Glover, G; Rivera, F; Mila, M. Incidence of fragile X in 5,000 consecutive newborn males. *Genet Test* 2003;7:339-343.

[7] Sherman, SL; Jacobs, PA; Morton, NE; Froster-Iskenius, U; Howard-Peebles, PN; Nielsen, KB; Partington, MW; Sutherland, GR; Turner, G; Watson, M. Further segregation analysis of the fragile X syndrome with special reference to transmitting males. *Hum Genet.* 1985;69:289-299.

[8] Mandel, JL & Biancalana, V. Fragile X mental retardation syndrome: from pathogenesis to diagnostic issues. *Growth Horm IGF Res* 2004;14:S158-165.

[9] Devys, D; Lutz, Y; Rouyer, N; Bellocq, JP; Mandel, JL. The FMR-1 protein is cytoplasmic, most abundant in neurons and appears normal in carriers of a fragile X premutation. *Nat Genet* 1993;4:335-340.

[10] Feng, Y; Absher, D; Eberhart, DE; Brown, V; Malter, HE; Warren, ST. FMRP associates with polyribosomes as an mRNP, and the I304N mutation of severe fragile X syndrome abolishes this association. *Mol Cell* 1997;1:109-118.

[11] Tamanini, F; Willemsen, R; van Unen, L; Bontekoe, C; Galjaard, H; Oostra, BA; Hoogeveen AT. Differential expression of FMR1, FXR1 and FXR2 proteins in human brain and testis. *Hum Mol Genet* 1997;6:1315-1322.

[12] Willemsen, R; Oostra, BA; Bassell, GJ; Dictenberg, J. The fragile X syndrome: from molecular genetics to neurobiology. *Ment Retard Dev Disabil Res Rev* 2004;10:60-67.

[13] Ashley, CT Jr; Wilkinson, KD; Reines, D; Warren, ST. FMR1 protein: conserved RNP family domains and selective RNA binding. *Science* 1993;262:563-566.

[14] Siomi, H; Siomi, MC; Nussbaum, RL; Dreyfuss, G. The protein product of the fragile X gene, FMR1, has characteristics of an RNA-binding protein. *Cell* 1993;74:291-298.

[15] Kiebler, MA & DesGroseillers, L. Molecular insights into mRNA transport and local translation in the mammalian nervous system. *Neuron* 2000;25:19-28.

[16] Steward, O. mRNA at synapses, synaptic plasticity, and memory consolidation. *Neuron* 2002;36:338-340.

[17] Comery, TA; Harris, JB; Willems, PJ; Oostra, BA; Irwin, SA; Weiler, IJ; Greenough, WT. Abnormal dendritic spines in fragile X knockout mice: maturation and pruning deficits. *Proc Natl Acad Sci U S A.* 1997;94:5401-5404.

[18] Reiss, AL; Freund, L; Abrams, MT; Boehm, C; Kazazian, H. Neurobehavioral effects of the fragile X premutation in adult women: a controlled study. *Am J Hum Genet* 1993;52:884-894.

[19] Brainard, SS; Schreiner, RA; Hagerman, RJ. Cognitive profiles of the carrier fragile X woman. *Am J Med Genet* 1991;38:505-508.

[20] Riddle, JE; Cheema, A; Sobesky, WE; Gardner, SC; Taylor, AK; Pennington, BF; Hagerman, RJ. Phenotypic involvement in females with the FMR1 gene mutation. *Am J Ment Retard* 1998;102:590-601.

[21] Bennetto, L; Pennington, BF; Porter, D; Taylor, AK; Hagerman, RJ. Profile of cognitive functioning in women with the fragile X mutation. *Neuropsychology* 2001;15:290-299.

[22] Hagerman, RJ. Physical and behavioral phenotype. In Hagerman RJ and Hagerman PJ, editors. *Fragile X Syndrome: Diagnosis, treatment and research*, 3rd ed. Baltimore, MD: The Johns Hopkins University Press. 2002.

[23] Hagerman, RJ; Staley, LW; O'Conner, R; Lugenbeel, K; Nelson, D; McLean, SD; Taylor, A. Learning-disabled males with a fragile X CGG expansion in the upper premutation size range. *Pediatrics* 1996;97:122-126.

[24] Hagerman, PJ & Hagerman, RJ. The fragile-X premutation: a maturing perspective. *Am J Hum Genet* 2004;74:805-816.

[25] Conway, GS. Premature ovarian failure. *Br Med Bull* 2000;56:643-649.

[26] Goswami, D & Conway GS. Premature ovarian failure. *Hum Reprod Update* 2005 11:391-410.

[27] Snowdon, DA; Kane, RL; Beeson, WL; Burke, GL; Sprafka, JM; Potter, J; Iso, H; Jacobs, DR Jr; Phillips, RL. Is early natural menopause a biologic marker of health and aging? *Am J Public Health* 1989;79:709-714.

[28] Conway, GS. Premature ovarian failure. *Curr Opin Obstet Gynecol* 1997;9:202-206.

[29] Sherman, SL. Premature ovarian failure in the fragile X syndrome. *Am J Med Genet* 2000;97:189-194.

[30] Cronister, A; Schreiner, R; Wittenberger, M; Amiri, K; Harris, K; Hagerman, RJ. Heterozygous fragile X female: historical, physical, cognitive, and cytogenetic features. *Am J Med Genet* 1991;38:269-274.

[31] Conway, GS; Hettiarachchi, S; Murray, A; Jacobs, PA. Fragile X premutations in familial premature ovarian failure. *Lancet* 1995;346:309-310.

[32] Conway, GS; Payne, NN; Webb, J; Murray, A; Jacobs, PA. Fragile X premutation screening in women with premature ovarian failure. *Hum Reprod* 1998;13:1184-1187.

[33] Vianna-Morgante, AM; Costa, SS; Pares, AS; Verreschi, IT. FRAXA premutation associated with premature ovarian failure. *Am J Med Genet* 1996;64:373-375.

[34] Murray, A; Webb, J; Grimley, S; Conway, G; Jacobs, P. Studies of FRAXA and FRAXE in women with premature ovarian failure. *J Med Genet* 1998;35:637-640.

[35] Mallolas, J; Duran, M; Sanchez, A; Jimenez, D; Castellvi-Bel, S; Rife, M; Mila, M. Implications of the FMR1 gene in menopause: study of 147 Spanish women. *Menopause* 2001;8:106-110.

[36] Allingham-Hawkins, DJ; Babul-Hirji, R; Chitayat, D; Holden, JJ; Yang, KT; Lee, C; Hudson, R; Gorwill, H; Nolin, SL; Glicksman; A; Jenkins, EC; Brown, WT; Howard-Peebles, PN; Becchi, C; Cummings, E; Fallon, L; Seitz, S; Black, SH; Vianna-Morgante, AM; Costa, SS; Otto, PA; Mingroni-Netto, RC; Murray, A; Webb, J; Vieri, F; *et al.* Fragile X premutation is a significant risk factor for premature ovarian failure: the International Collaborative POF in Fragile X study--preliminary data. *Am J Med Genet* 1999; 83:322-325.

[37] Hundscheid, RD; Braat, DD; Kiemeney, LA; Smits, AP; Thomas, CM. Increased serum FSH in female fragile X premutation carriers with either regular menstrual cycles or on oral contraceptives. *Hum Reprod* 2001;16:457-462.
[38] Hundscheid, RD; Sistermans, EA; Thomas, CM; Braat, DD; Straatman, H; Kiemeney, LA; Oostra, BA; Smits, AP. Imprinting effect in premature ovarian failure confined to paternally inherited fragile X premutations. *Am J Hum Genet* 2000;66:413-418.
[39] Vianna-Morgante, AM & Costa, SS. Premature ovarian failure is associated with maternally and paternally inherited premutation in Brazilian families with fragile X. *Am J Hum Genet* 2000;67:254-255.
[40] Murray, A; Ennis, S; Morton, N. No evidence for parent of origin influencing premature ovarian failure in fragile X premutation carriers. *Am J Hum Genet* 2000;67:253-254.
[41] Sherman, SL. ature ovarian failure among fragile X premutation carriers: parent-of-origin effect? *Am J Hum Genet* 2000;67:11-13.
[42] Murray, A; Ennis, S; MacSwiney, F; Webb, J; Morton, NE. Reproductive and menstrual history of females with fragile X expansions. Eur J Hum Genet 2000;8:247-252.
[43] Rife, M; Nadal, A; Mila, M; Willemsen, R. Immunohistochemical FMRP studies in a full mutated female fetus. *Am J Med Gene* 2004;124:129-132.
[44] Bachner, D; Manca, A; Steinbach, P; Wohrle, D; Just, W; Vogel, W; Hameister, H; Poustka, A. Enhanced expression of the murine FMR1 gene during germ cell proliferation suggests a special function in both the male and the female gonad. *Hum Mol Genet* 1993;2:2043-2050.
[45] Hagerman, RJ; Leehey, M; Heinrichs, W; Tassone, F; Wilson, R; Hills, J; Grigsby, J; Gage, B; Hagerman, PJ. Intention tremor, parkinsonism, and generalized brain atrophy in male carriers of fragile X. *Neurology* 2001;57:127-130.
[46] Jacquemont, S; Hagerman, RJ; Leehey, MA; Hall, DA; Levine, RA; Brunberg, JA; Zhang, L; Jardini, T; Gane, LW; Harris, SW; Herman, K; Grigsby, J; Greco, CM; Berry-Kravis, E; Tassone, F; Hagerman, PJ. Penetrance of the fragile X-associated tremor/ataxia syndrome in a premutation carrier population. *JAMA* 2004;291:460-469.
[47] Brunberg, JA; Jacquemont, S; Hagerman, RJ; Berry-Kravis, EM; Grigsby, J; Leehey, MA; Tassone, F; Brown, WT; Greco, CM; Hagerman, PJ. Fragile X premutation carriers: characteristic MR imaging findings of adult male patients with progressive cerebellar and cognitive dysfunction. *AJNR Am J Neuroradiol* 2002;23:1757-1766.
[48] Jacquemont, S; Hagerman, RJ; Leehey, M; Grigsby, J; Zhang, L; Brunberg, JA; Greco, C; Des Portes, V; Jardini, T; Levine, R; Berry-Kravis, E; Brown, WT; Schaeffer, S; Kissel J; Tassone, F; Hagerman, PJ. Fragile X premutation tremor/ataxia syndrome: molecular, clinical, and neuroimaging correlates. *Am J Hum Genet* 2003;72:869-878.
[49] Jacquemont, S; Farzin, F; Hall, D; Leehey, M; Tassone, F; Gane, L; Zhang, L; Grigsby, J; Jardini, T; Lewin, F; Berry-Kravis, E; Hagerman, PJ; Hagerman, RJ. Aging in individuals with the FMR1 mutation. *Am J Ment Retard* 2004;109:154-164.
[50] Macpherson, J; Waghorn, A; Hammans, S; Jacobs, P. Observation of an excess of fragile-X premutations in a population of males referred with spinocerebellar ataxia. *Hum Genet* 2003;112:619-620.

[51] Zuhlke, Ch; Budnik, A; Gehlken, U; Dalski, A; Purmann, S; Naumann, M; Schmidt, M; Burk, K; Schwinger, E. FMR1 premutation as a rare cause of late onset ataxia--evidence for FXTAS in female carriers. *J Neurol* 2004;251:1418-1419

[52] Brussino, A; Gellera, C; Saluto, A; Mariotti, C; Arduino, C; Castellotti, B; Camerlingo, M; de Angelis, V; Orsi, L; Tosca, P; Migone, N; Taroni, F; Brusco, A. FMR1 gene premutation is a frequent genetic cause of late-onset sporadic cerebellar ataxia. *Neurology* 2005;64:145-147.

[53] Van Esch, H; Matthijs, G; Fryns, JP. Should we screen for FMR1 premutations in female subjects presenting with ataxia? *Ann Neurol* 2005;57:932-933.

[54] Tassone, F; Hagerman, RJ; Taylor, AK; Gane, LW; Godfrey, TE; Hagerman, PJ. Elevated levels of FMR1 mRNA in carrier males: a new mechanism of involvement in the fragile-X syndrome. *Am J Hum Genet* 2000;66:6-15.

[55] Tassone, F; Hagerman, RJ; Taylor, AK; Mills, JB; Harris, SW; Gane, LW; Hagerman, PJ. Clinical involvement and protein expression in individuals with the FMR1 premutation. *Am J Med Genet* 2000;91:144-152.

[56] Kenneson, A; Zhang, F; Hagedorn, CH; Warren, ST. Reduced FMRP and increased FMR1 transcription is proportionally associated with CGG repeat number in intermediate-length and premutation carriers. *Hum Mol Genet* 2001;10:1449-1454.

[57] Oostra, BA & Willemsen, R. A fragile balance: FMR1 expression levels. *Hum Mol Genet* 2003;12:R249-R257.

[58] Greco, CM; Hagerman, RJ; Tassone, F; Chudley, AE; Del Bigio, MR; Jacquemont, S; Leehey, M; Hagerman, PJ. Neuronal intranuclear inclusions in a new cerebellar tremor/ataxia syndrome among fragile X carriers. *Brain* 2002;125:1760-1771.

[59] Mankodi, A & Thornton, CA. Myotonic syndromes. *Curr Opin Neurol* 2002;15:545-552.

[60] Mankodi, A; Logigian, E; Callahan, L; McClain, C; White, R; Henderson, D; Krym, M; Thornton, CA. Myotonic dystrophy in transgenic mice expressing an expanded CUG repeat. *Science* 2000;289:1769-1773.

[61] Greco, CM; Berman, RF; Martin, RM; Tassone, F; Schwartz, PH; Chang, A; Trapp, BD; Iwahashi, C; Brunberg, J; Grigsby, J; Hessl, D; Becker, EJ; Papazian, J; Leehey, MA; Hagerman, RJ; Hagerman, PJ. Neuropathology of fragile X-associated tremor/ataxia syndrome (FXTAS). *Brain* 2006;129:243-255.

[62] Bontekoe, CJ; Bakker, CE; Nieuwenhuizen, IM; van der Linde, H; Lans, H; de Lange, D; Hirst, MC; Oostra, BA. Instability of a (CGG)98 repeat in the Fmr1 promoter. *Hum Mol Genet.* 2001; 10: 1693-1639.

[63] Willemsen R, Hoogeveen-Westerveld M, Reis S, Holstege J, Severijnen LA, Nieuwenhuizen IM, Schrier M, van Unen L, Tassone F, Hoogeveen AT, Hagerman PJ, Mientjes EJ, Oostra BA. The FMR1 CGG repeat mouse displays ubiquitin-positive intranuclear neuronal inclusions; implications for the cerebellar tremor/ataxia syndrome. *Hum Mol Genet.* 2003; 12: 949-959.

[64] Loesch DZ, Huggins R, Petrovic V, Slater H. Expansion of the CGG repeat in fragile X in the FMR1 gene depends on the sex of the offspring. *Am J Hum Genet.* 1995 Dec;57(6):1408-13.

[65] Nolin SL, Lewis FA 3rd, Ye LL, Houck GE Jr, Glicksman AE, Limprasert P, Li SY, Zhong N, Ashley AE, Feingold E, Sherman SL, Brown WT. Familial transmission of the FMR1 CGG repeat. *Am J Hum Genet.* 1996 Dec;59(6):1252-61.

[66] McConkie-Rosell A, Finucane B, Cronister A, Abrams L, Bennett RL, Pettersen BJ. Genetic counseling for fragile x syndrome: updated recommendations of the national society of genetic counselors. *J Genet Couns*. 2005 Aug;14(4):249-70.

[67] de Vries BB, Wiegers AM, Smits AP, Mohkamsing S, Duivenvoorden HJ, Fryns JP, Curfs LM, Halley DJ, Oostra BA, van den Ouweland AM, Niermeijer MF. Mental status of females with an FMR1 gene full mutation. *Am J Hum Genet* 1996;58:1025-1032.

[68] Rousseau, F; Heitz, D; Biancalana, V; Blumenfeld, S; Kretz, C; Boue, J; Tommerup, N; Van Der Hagen, C; DeLozier-Blanchet, C; Croquette, MF; *et al.* Direct diagnosis by DNA analysis of the fragile X syndrome of mental retardation. *N Engl J Med* 1991;325:1673-1681.

[69] Taylor, AK; Safanda, JF; Fall, MZ; Quince, C; Lang, KA; Hull, CE, Carpenter, I; Staley, LW; Hagerman, RJ. Molecular predictors of cognitive involvement in female carriers of fragile X syndrome. *JAMA* 1994;271:507-514.

[70] Malter, HE; Iber, JC; Willemsen, R; de Graaff, E; Tarleton, JC; Leisti, J; Warren, ST; Oostra, BA. Characterization of the full fragile X syndrome mutation in fetal gametes. *Nat Genet* 1997;15:165-169.

[71] Wohrle, D; Hennig, I; Vogel, W; Steinbach, P. Mitotic stability of fragile X mutations in differentiated cells indicates early post-conceptional trinucleotide repeat expansion. *Nat Genet* 1993;4:140-142.

[72] Kruyer, H; Mila, M; Glover, G; Carbonell, P; Ballesta, F; Estivill, X. Fragile X syndrome and the (CGG)n mutation: two families with discordant MZ twins. *Am J Hum Genet* 1994;54:437-442.

[73] Nolin, SL; Brown, WT, Glicksman, A; Houck, GE Jr; Gargano, AD; Sullivan, A; Biancalana, V; Brondum-Nielsen, K; Hjalgrim, H; Holinski-Feder, E; Kooy, F; Longshore, J; Macpherson, J; Mandel, JL; Matthijs, G; Rousseau, F; Steinbach, P; Vaisanen, ML; von Koskull, H; Sherman, SL. Expansion of the fragile X CGG repeat in females with premutation or intermediate alleles. *Am J Hum Genet* 2003;72:454-464.

[74] Rife, M; Badenas, C; Quinto, L; Puigoriol, E; Tazon, B; Rodriguez-Revenga, L; Jimenez, L; Sanchez, A; Mila, M. Analysis of CGG variation through 642 meioses in Fragile X families. *Mol Hum Reprod* 2004;10:773-776.

[75] Reyniers, E; Vits, L; De Boulle, K; Van Roy, B; Van Velzen, D; de Graaff, E; Verkerk, AJ; Jorens, HZ; Darby, JK; Oostra, B; *et al.* The full mutation in the FMR-1 gene of male fragile X patients is absent in their sperm. *Nat Genet* 1993;4:143-146.

[76] Welt, CK; Smith, PC; Taylor, AE. Evidence of early ovarian aging in fragile X premutation carriers. *J Clin Endocrinol Metab* 2004;89:4569-4574.

[77] Castellvi-Bel, S; Mila, M; Soler, A; Carrio, A; Sanchez, A; Villa, M; Jimenez, MD; Estivill, X. Prenatal diagnosis of fragile X syndrome: (CGG)n expansion and methylation of chorionic villus samples. *Prenat Diagn* 1995;15:801-807.

[78] Rodriguez-Revenga, L; Badenas, C; Madrigal, I; Sanchez, A; Soler, A; Carrio, A; Mila, M. 46,XY,18q+/46,XY,18q- mosaicism in a fragile X prenatal diagnosis. *Prenat Diagn* 2005;25:448-450.

[79] Platteau, P; Sermon, K; Seneca, S; Van Steirteghem, A; Devroey, P; Liebaers, I. Preimplantation genetic diagnosis for fragile Xa syndrome: difficult but not impossible. *Hum Reprod* 2002;17:2807-2812.

[80] Fu, YH; Kuhl, DP; Pizzuti, A; Pieretti, M; Sutcliffe, JS; Richards, S; Verkerk, AJ; Holden, JJ; Fenwick, RG Jr; Warren, ST; *et al*. Variation of the CGG repeat at the fragile X site results in genetic instability: resolution of the Sherman paradox. *Cell* 1991;67:1047-1058.

[81] Rife, M; Mallolas, J; Badenas, C; Tazon, B; Miguelez, MR; Pampols, T; Sanchez, A; Mila, M. Pilot study for the neonatal screening of fragile X syndrome. *Prenat Diagn* 2002;22:459-462.

[82] Rousseau, F; Rouillard, P; Morel, ML; Khandjian, EW; Morgan, K. Prevalence of carriers of premutation-size alleles of the FMRI gene--and implications for the population genetics of the fragile X syndrome. *Am J Hum Genet* 1995;57:1006-1018.

[83] Anido, A; Carlson, LM; Taft, L; Sherman, SL. Women's attitudes toward testing for fragile X carrier status: a qualitative analysis. *J Genet Couns* 2005;14:295-306.

In: New Research on Fragile X Syndrome
Editor: Jeffrey J. Upner, pp. 117-134
ISBN: 1-60021-281-6

Chapter 7

SEARCHING FOR COMMON GROUND: THE FRAGILE X MENTAL RETARDATION PROTEIN AS A CORE ELEMENT OF RNA GRANULES

***Ying Ju Sung*[1]*, Natalia Dolzhanskaya*[2] *and Robert B. Denman*[2]**
[1]Columbia University, USA, 630 West 168th Street, New York, NY 10032
[2]New York State Institute for Basic Research, In Developmental Disabilities, USA, 1050 Forest Hill Road, Staten Island, NY 10314

ABSTRACT

The fragile X mental retardation protein FMRP is an RNA binding protein that associates with a host of mRNAs and with other cellular RNA binding proteins to form messenger ribonucleoprotein particles (mRNPs). These mRNPs, in turn, associate with ribosomes and it has been shown that FMRP selectively represses the translation of some of the messages to which it binds. In addition, FMRP is a component of a variety of supramolecular RNA/protein granules including: the RNA induced silencing complex (RISC) complex, stress granules, transport granules and mRNA granules. We have recently identified a unique and smaller cytoplasmic FMRP granule. The nature of this granule is not fully characterized, yet it is known that its presence increases when cellular methylation is blocked. Here, we demonstrate that small cytoplasmic FMRP granules are a common feature of cultured cell lines. We also show by *in situ* hybridization to an $Alexa_{488}$-oligo dT_{18} probe that this granule is devoid of poly A^+ RNA. Furthermore, in cells treated with the methylation inhibitor AdOx prior to the induction of stress granules many FMRP-granules are physically associated with, albeit distinct from, stress granules. These and other circumstantial data suggest that the small cytoplasmic FMRP granules found in cultured cells might be mRNA processing bodies (P-bodies). Endogenous FMRP is recruited into abortive pre-initiation complexes (48S*) called stress granules following heat shock, treatment with sodium arsenite and mechanical injury. These granules are the major repository for FMRP following oxidative stress and we show that while FMRP and the other fragile X-related proteins (FXRPs) form stress granules, not all proteins do so. This indicates that although heterogeneous in size, stress granules have a limited composition. We also report that the neurites of arsenite-treated PC12 cells harbor a heterogeneous population of FMRP granules. Some co-localize with the T cell internal antigen 1 (TIA-1), an RNA binding protein that defines classic stress granules. Some co-localize with FXR1P or FXR2P. However, in each instance the co-localization was incomplete. This is in contrast to the arsenite-induced FMRP granule populations in the cell soma and in the growth cones, which by-and-large uniformly contain all four proteins. Our data support a model in which sodium arsenite treatment provides a snapshot of granule trafficking by marking potentially translational mRNA granules from those that have delivered their messages and are transiting back to the nucleus for others. Thus, FMRP lies at the confluence of many types of granules and the underlying biology of these granules may provide the molecular basis for the wide-ranging phenotype seen in fragile X syndrome.

INTRODUCTION

In the "*ome-ish*" world that is today's science, large-scale interactions and supra-molecular complexes are the norm. For example, an integrated network combining synthetic sick or lethal (SSL) interactions, protein sequence homology, correlated mRNA expression data, stable protein-protein interactions and transcriptional regulatory interactions has led to the construction of a yeast interactome consisting of 154,739 links [67]. This six-degrees-of-separation-view of the cellular landscape at once, provides a unique glimpse into global cellular metabolism, but challenges our basic reductionist assumptions concerning the underlying building blocks of cells and how these building blocks carry out the biochemical reactions necessary for life. Of course, precedence for this view had existed for decades in the form of well-characterized assemblages such as the ribosome, the spliceosome and the fatty acid synthesis complex, but the current list of interacting protein complexes dwarfs these harbingers exponentially.

One example of how "*omics*" has impacted our understanding of cellular metabolism is in the field of RNA binding proteins. Traditional biochemistry parsed the RNA binding proteins into several categories. In addition, to large protein/RNA complexes such as spliceosomes and ribosomes, which carry out defined reactions on pre-messenger and messenger RNAs, smaller complexes *ca.* 660 KDa called messenger ribonucleoprotein particles (mRNPs) were also listed as discretely functioning entities. mRNPs consist of a defined message and a small cohort of accessory proteins that associate directly with specific motifs within the message (RNA binding proteins such as the polyadenylation binding protein (PABP) or the cap binding protein) or indirectly through association with other proteins (maskin) [6]. As a basic unit of cellular currency, mRNPs have been shown to be involved in mRNA localization, mRNA stability and translational repression [24; 31; 49; 51]. In addition, mRNPs were viewed as dynamic components as they could respond to specific cellular cues [40]. However, the discovery and characterization of large RNA granules has changed this picture considerably. RNA granules are supramolecular complexes of proteins and RNAs. To date, four types have been at least partially characterized. They include: stress granules, P-bodies, mRNA transport granules and the RISC complex, although others no doubt exist.

Stress Granules

Oxidative stress and heat shock produces a well-characterized response in non-neuronal cells, which mobilizes the translational machinery to cope with the insult. Specifically, mRNAs normally translated by the cells are shunted into non-translated messenger ribonucleoprotein (mRNP) granules *i.e.* stress granules, while heat shock mRNAs are selectively translated [27]. The mechanism of stress granule formation is well established [1]. Phosphorylation of eukaryotic initiation factor 2-alpha (eIF2α) by one of several kinases inhibits the formation of a ternary complex between itself, initiator Met-tRNA and GTP. This, in turn, results in (1) the translocation of the T-cell internal antigen (TIA-1) from the nucleus to the cytoplasm and (2) its subsequent recruitment, into 48S* pre-initiation complexes, which are then no longer competent to form 80S monosomes. Further addition of TIA-1 results in the aggregation of individual 48S* pre-initiation complexes into stress granules. Additionally, stress granule formation may also occur when certain RNA binding proteins are heterologously over expressed. Examples of this are the RasGAP-associated endoribonuclease (G3BP), the survival of motor neuron protein (SMN), the cytoplasmic polyadenylation binding protein 1 (CPEB1) and Disrupted-In-Schiophrenia 1 (DISC1) [23; 46; 60; 65].

P-Bodies

mRNA degradation occurs sequentially. In the 5'-3' pathway, a message is first deadenylated. Next, the methyl-G-cap is removed. The uncapped message is then susceptible attack by 5' to 3' ribonucleases, leading to its degradation [20]. Cap removal requires the mRNA to transit from polyribosomes and into a repressed mRNP complex that lacks translation initiation factors and contains the decapping enzymes dcp1a and dcp2, the endoribonuclease Xrn1, and several other proteins including Ge-1, Sm-like proteins 1–7, the DEAD box family helicase Rck/p54 and the autoantigen GW182 [28; 66]. The aggregation of repressed mRNPs results in the formation of P-bodies [59], which are also called GW or Dcp bodies [42]. Recent studies have revealed that stress granules and P-bodies are dynamically linked [65], such that proteins that are common to both transit between the two granules at variable rates [28].

Transport Granules

Polarized cells such as neurons use local protein synthesis to respond rapidly to stimuli far from the cell soma. Thus, all of the components for protein translation: ribosomes, initiation factors, elongation factors, mRNAs, and regulatory RNAs and proteins have been detected in dendrites [43; 54; 57; 64]. Furthermore, protein synthesis of endogenous and reporter mRNAs have been demonstrated in dendrites that have been detached from their cell bodies [9; 19]. These factors are shipped from the soma *via* the microtuble network by kinesin motors in the form of translationally silent granules [30]. These granules contain proteins, rRNA and mRNA, but lack tRNA and certain other ribosomal factors.

The RISC Complex

The RNA induced silencing complex (RISC), a ribonucleoprotein particle, mediates mRNA degradation *via* siRNA-induced cleavage and translational repression *via* miRNA-induced protein synthesis inhibition. In reality, RISC may be a heterogeneous population of particles differing in both their RNAs (siRNA *vs.* miRNA) and proteins. A variety of proteins have been associated with RISC including Gemin 3, Gemin 4, FMRP, FXR1P, Dmp68 Tsn, Vig and the argonaute proteins 1-4 (Ago-1-4); however, only Ago2 has been detected in all RISCs isolated to date [53]. These data coincide with the observation that siRNA-directed mRNA cleavage activity has been reported to reside in complexes of ~160 kDa, ~200 kDa, ~500 kDa, ~550 kDa and ~80S making it likely that there is more than one RISC. Indeed, some of the smaller complexes appear to be intermediates in the formation of the 80S holo-RISC [47].

The Fragile X Mental Retardation Protein, FMRP

The fragile X mental retardation protein (FMRP) is a prototypical RNA binding protein that lies at the confluence of RNA granule studies. The loss of FMRP, as occurs in fragile X syndrome (FXS), results in a large number of phenotypic features ranging from mild-severe cognitive deficits (IQs of 75 or less), macroorchidism, altered skeletal features (large ears, elongated face, weak and hyper extendible joints) as well as debilitating behavioral abnormalities (attention deficit hyperactivity disorder (ADHD), autism and seizures) [16; 21; 44; 48]. The ability of FMRP to bind RNA [3; 52] led naturally to the hypothesis that the wide ranging phenotype associated with this syndrome stems from altered processing, stability, transport or translation of the RNAs to which it binds [17]. Thus, small and large scale genomic and proteomic studies have been carried out to identify RNAs and proteins that interact with FMRP. One of the hypotheses that rose from this work implied that FMRP might exist as part of a larger complex [12] and this has been borne out. Indeed, FMRP has been found to be a component of stress granules, transport granules and the RISC complex and much about what we know about these particles can be attributed to studies aimed at understanding FMRP localization and function. In this chapter the results of new FMRP granules studies will be set within the context of previous results in the hope that the diversity of FMRP RNA granule function will be better appreciated.

RESULTS

Small Cytoplasmic FMRP Granules are a Common Feature of Cultured Cell Lines

We previously documented the presence of endogenous FMRP-containing small cytoplasmic granules in HeLa cells by immunostaining with a commercially available antibody directed to an epitope in the N-terminal 200 amino acids of FMRP [13; 14]. These granules, which were observed in about 30% of all HeLa cells, were estimated to be 0.5-1.5 μm^3 by confocal imaging. Furthermore, the granules were present in both mitotic and non-mitotic cells. Since the existence of such particles had not been previously reported we probed several commonly used cultured cell lines with the anti-FMRP antibody to determine whether this feature was unique to HeLa cell FMRP, or was a more general property. The results of this study are shown in Figure 1. Indeed, it was observed that these granules were also present to varying degrees in human rhabdomyosarcoma cells (A204), rat pheochromocytoma (PC12) cells and human lymphoblastoid cells. Thus, it would appear that a subpopulation of FMRP normally exists in cells as small cytoplasmic granules.

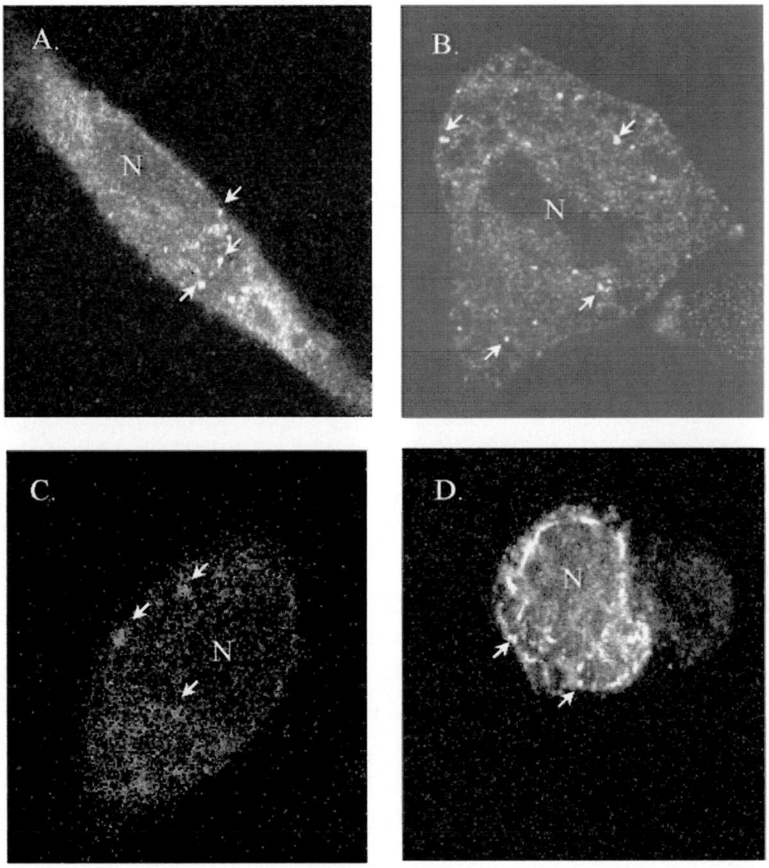

Figure 1. FMRP cytoplasmic granules are a common feature of cultured cell lines. (A) Rhabdomyosarcoma A204 cells, (B) HeLa cells, (C) PC12 cells, (D) lymphoblastoid cells. Arrows mark cytoplasmic granules, N marks the nucleus. Cells grown on glass coverslips were fixed in 2% paraformaldehyde for ten minutes and washed with 1xPBS and then blocked in (RPMI1640 base medium, 0.05% saponin, 0.1% sodium azide, 2% goat serum) for 30 min. at room temperature. For immunostaining the coverslips were equilibrated in 1xPBS and 0.1% Triton X-100 (PBST), blocked with 10% mouse serum in PBST for 30 min at room temperature, and then incubated with anti-FMRP antibody for 1 hour. This was followed by a 30 min incubation with Alexa $Fluor_{568}$-conjugated secondary antibody (red) (1:500 dilution) at room temperature. Finally, the coverslips were washed in RPMI 1640 base medium, 1% goat serum, and mounted in buffered glycerol. Fluorescence was detected with a NIKON PCM 2000 dual laser scanning confocal microscope (Nikon). Images were acquired at a magnification of 60x.

Small Cytoplasmic FMRP Granules are Devoid of Poly A^+ RNA

Both transport granules and stress granules are known to contain mRNAs. Since FMRP has previously been shown to be a component of these two types of granules [25; 38] it was of interest to determine whether the small cytoplasmic FMRP granules we observed in cultured cells (above) (1) were stress granules or (2) also harbored mRNA. To address the first question HeLa cells were immunostained with anti-FMRP and known stress granule marker protein antibodies. The results conclusively demonstrated that the small cytoplasmic FMRP granules were not stress granules [14], indicating that under the conditions we grew the cells they were not under stress. To address the second question HeLa cells were probed concurrently with an $Alexa_{488}$-oligo dT_{18} probe using an established *in situ* hybridization protocol to image poly A^+ RNA [56] and anti-FMRP antibody to visualize FMRP granules. Representative results of these studies are presented in Figure 2A. FMRP immunostaining was localized primarily to the cytoplasm as previously shown. Again, a fraction of the cells had discernable granules. In contrast, a significant fraction of the $Alexa_{488}$-oligo dT_{18} probe fluorescence was detected in the nucleus, concentrated in large amorphous structures that are reminiscent of active zones of transcription. $Alexa_{488}$-oligo dT_{18} probe fluorescence was also found in the cytoplasm; however, there was never any correspondence with the FMRP cytoplasmic granules. Thus, the data are compatible with the hypothesis that the FMRP small cytoplasmic granules do not contain poly A^+ RNA.

Several possibilities may account for this result. It is important to note however, that technical issues with the method are not responsible. Thus, while the fluorescence intensity of the *in situ* probe was weak, it could be clearly detected throughout the cytoplasm, and in instances where FMRP granules were present there was no localized increase in the staining intensity, Figure 2B. In contrast, Mazroui *et al* and Antar *et al* have localized RNA in stress granules, and transport granules, respectively, and in the former studies the *in situ* probe was similar to the one used here [2; 38]. Another "quasi-technical" explanation of the results is that the poly A^+ tail of the mRNA is masked within the FMRP cytoplasmic granules and thus, not available for hybridization. In favor of this, Dolzhanskaya *et al* have demonstrated that FMRP immunoprecipitates (IPs) prepared from cells that were treated or not treated with the general methylation inhibitor adenosine 2'-3'-dialdehyde (AdOx), which significantly increases FMRP small cytoplasmic granules, both contain FMRP target mRNAs; however, the absolute amounts differ [14]. On the other hand, this result may be equally well explained if the mRNA in these granules were deadenylated, or absent altogether. The former result is compatible with data showing that in cells treated with AdOx prior to stress granule induction, some of the smaller FMRP granules are associated with stress granules as P-bodies are expected to be, Figure 2C. In addition, Liu *et al* have recently shown that Argonaute proteins (Ago) of the RNA induced silencing complex (RISC) are found in small cytoplasmic foci in U2-OS cells [35]. By virtue of their co-localization with decapping proteins dcp1 and dcp2, these foci are P-bodies. The number, size and distribution of the Ago-foci are very much like the FMRP cytoplasmic granules shown in Figure 1. Furthermore, FMRP has been found in RISC [7; 36], making this a very attractive possibility. Future studies that determine whether these granules are, in fact, mRNA processing bodies will be needed to fully resolve this question.

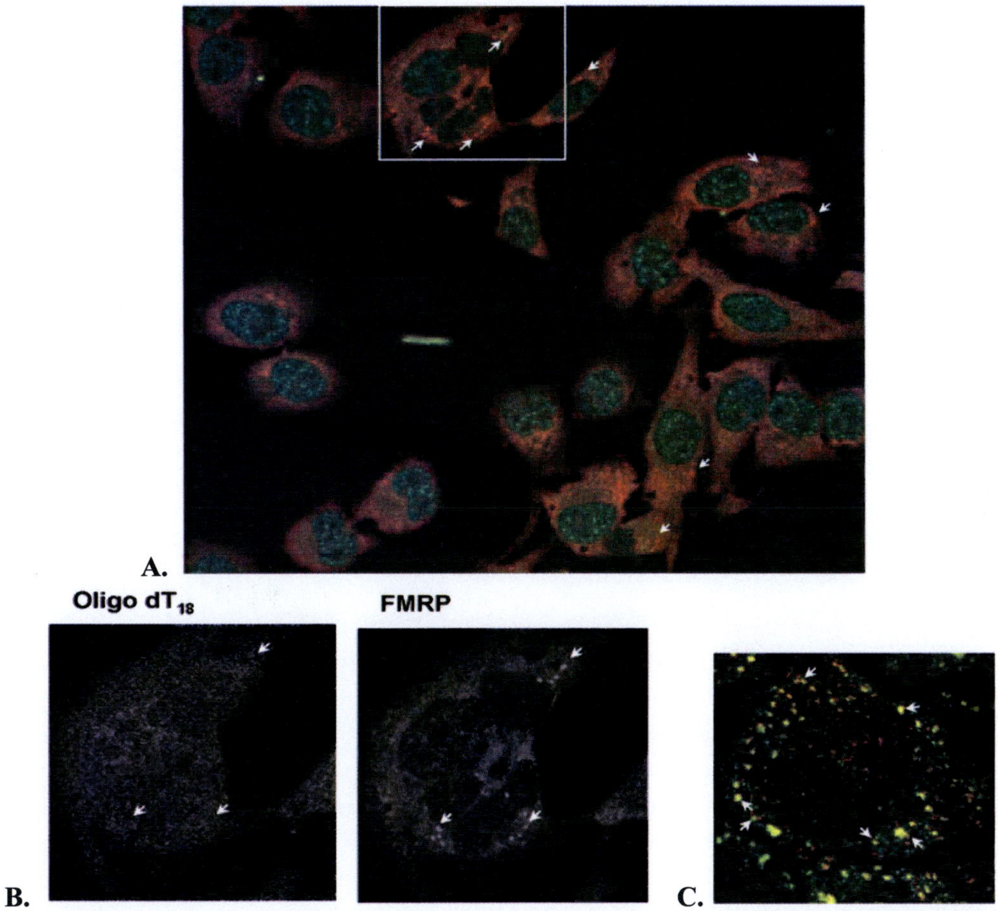

Figure 2. FMRP cytoplasmic granules do not co-localize with poly A^+ RNA. (A) HeLa cells, grown on glass coverslips were first fixed in 4% paraformaldehyde in PBS, pH 7.4, for 3 hr. The coverslips were subsequently washed several times in 1xPBS and then treated with 80 μg/ml proteinase K in 1xPBS for 30 min at room temperature. After several washes in 1xPBS, the cells were fixed again for 20 min with 4% paraformaldehyde and washed again with 1xPBS. After a 10 min incubation with 1.32% triethanolamine HCl, pH 8.0, and a 20 min treatment with 0.24% acetic anhydride in 1.32% triethanolamine HCl, pH 8.0 the cells were washed with 1xPBS and then pre-hybridized in 50% formamide, 5xSSC, 5xDenhardt's solution, 0.25 mg/ml yeast tRNA, and 0.5 mg/ml salmon sperm DNA at 60C for 2 hr. Hybridization was performed overnight at 42°C with fresh buffer containing an $Alexa_{488}$-oligo dT_{18} (green) probe. After hybridization, the cells were washed for 30 min in fresh buffer at 42C, and then in 0.2xSSC at 42C for 1 hr. For immunostaining the hybridized coverslips were equilibrated in 1xPBS and 0.1% Triton X-100 (PBST), blocked with 10% mouse serum in PBST for 30 min at room temperature, and then incubated with anti-FMRP antibody (red) for 1 hour and processed as described in Figure 1. Arrows show typical FMRP cytoplasmic granules without co-localizing poly A^+ RNA. (B) $Alexa_{488}$-oligo dT_{18} fluorescence and FMRP immunofluorescence of the boxed area in (A), showing uniformly diffuse but clear hybridization of the $Alexa_{488}$-oligo dT_{18} probe within the cytoplasm, especially near FMRP granules (arrows). (C) FMRP small cytoplasmic granules are often associated with stress granules after treatment with AdOx. HeLa cells were pre-treated for 24 hr with 20 μM AdOx. Subsequently, they were treated with 0.5 mM sodium arsenite and fixed and immunostained with anti-FMRP (red) and anti-PABP (green) antibodies. Arrows show instances where smaller FMRP granules associate, but are distinct from the larger (green and yellow) stress granules, a feature that is reminiscent of P-bodies.

Finally, the latter possibility (absence of RNA) is consistent with the notion that such granules may be formed from incompletely post-translationally modified forms of FMRP *i.e.* immature products that can't interact with their normal complement of mRNA.In support of this Dolzhanskaya *et al* found that an antibody, which recognized methylated FMRP on Western blots, did not immunostain these granules [14]. Biochemical characterization studies must definitively establish whether these granules are methylated or phosphorylated.

FMRP is a Component of Stress Granules

As mentioned in the introduction to this chapter, there are several routes to stress granule formation. These include environmental insults such as heat shock, exposure to ultraviolet light, oxidative stress, cellular energy depletion, global ischemia and reperfusion [26; 39; 60] as well as artificial stimuli like the over expression of the RNA binding proteins that form these granules [23; 46; 60; 65] or the expression of a phospho-mimetic mutant form of eIF-2α [27]. In addition, recent work suggests that mechanical injury can also lead to stress granule formation [29].

Mazroui *et al* produced the first evidence that FMRP was associated with stress granules [38]. Indeed, they demonstrated that over expressing FMRP led to the formation of granules that also contained PABP, FXR1P and poly A^+ RNA, but did not contain ribosomal protein L7. Additionally, standard heat shock of HeLa cells (43C, 30 min) also led to granules with the same composition. Largely based on morphology and the absence of ribosomal protein L7, these granules were called stress granules. Recent studies by members of this laboratory have expanded this work by demonstrating that FMRP-containing stress granules form naturally following the treatment of HeLa cells [14] and PC12 cells [15] with sodium arsenite. These granules were identified based on co-localization with PABP and TIA-1. Finally, Kim *et al* have shown that injuries resulting from electrode insertion and decapitation result in a shift of FMRP from polyribosomes and into TIA-1-containing stress granules [29]. The conclusion to be drawn from all of these studies is that, unlike tristetraprolin (TTP) [58], FMRP is a core component of stress granules and its incorporation is independent of the pathway of induction.

TIA-1, another core component of stress granules, plays a defining role in determining their morphology. Specifically, a glutamine-rich prion-related domain (PRD) on the C-terminal end of the molecule is the primary factor in promoting the assembly of these granules, although other prion-like domains can artificially substitute [18]. Aggregation, gives stress granules their characteristic amorphous and highly heterogeneous size and shape. Figure 3A (upper panel) shows representative results of the FMRP-containing stress granules in HeLa cells induced by the application of sodium arsenite. It is readily apparent that not all of the FMRP and TIA-1 co-localized granules are of uniform size (lower panel). Nevertheless, the fact that their size and shape are not uniform does not mean that they do not have a limited composition. This is illustrated in the right hand panels of Figure 3B, which show the results of a co-localization study between TIA-1 and the constitutively expressed form of heat shock 70 protein (Hsp70cp), following arsenite treatment. In this case, despite the presence of large perinuclear TIA-1 granules, the majority of Hsp70cp remains in smaller nuclear puncta, and in instances where Hsp70cp granules are visible in the cytoplasm, they lie adjacent to the larger TIA1-containing granules (see insets below). The latter finding is consistent with the role Hsp70 plays in mRNA triage [1].

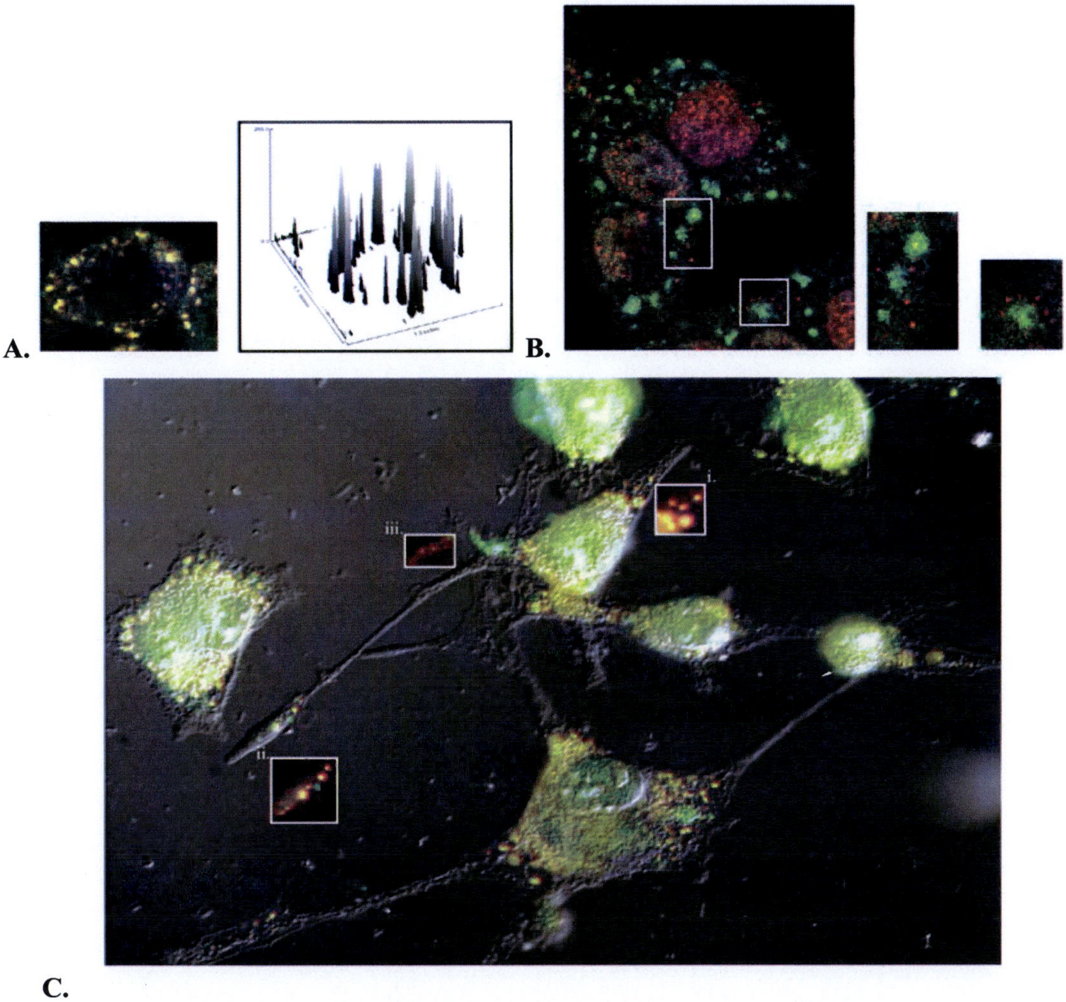

Figure 3. Characterization of FMRP-containing stress granules. (A) Confocal immunofluorescence image showing that FMRP (red) forms granules (yellow) that co-localize with the stress granule marker TIA-1 (green) in HeLa cells treated with 0.5 mM sodium arsenite for 30 min. (upper panel). The lower panel shows a surface plot of granule size (x-y axes) and intensity (z axis) within the cell above. (B) The constitutively expressed heat shock 70 protein Hsp70cp (red), however is found primarily in nuclear granules and to a lesser degree, small cytoplasmic granules adjacent to TIA-1 stress granules (green). The insets below the full image show magnified views of the boxed regions. Images were acquired as described in Figure 1. (C) Heterogeneity of FMRP-containing granules under oxidative stress conditions. PC12 cells were treated with NGF (200 ng/ml) for 48 hr to initiate the development of neurites. Subsequently, the cells were treated with 0.5 mM sodium arsenite for 20 min and then immunostained with anti-FMRP (red) and anti-TIA-1 (green). Composite immunofluorescence images were superimposed on a digital interference contrast image (DIC). Insets, which lack the DIC image, show (i) FMRP and TIA-1 granules in the soma, (ii) FMRP and TIA-1 granules in a growth cone and (iii) FMRP and TIA-1 granules in the neurites of a single differentiated PC12 cell.

While limited in their overall composition, stress granules are dynamic entities. For example, using fluorescence recovery after photobleaching (FRAP) Kedersha *et al* have shown that PABP and TIA-1 are not always in the same granules, while TIA-1 and G3BP appear to be [28]. Additional experiments by this group have demonstrated that proteins that are components of both stress granules and P-bodies (TTP, FAST) transit between the two granules at different rates.

Recent studies by members of this laboratory on the effect FMRP methylation plays in its function have led to similar conclusion. Adenosine 2'-3'-dialdehyde (AdOx) is a feedback inhibitor of protein arginine methylation and treating cells with micromolar concentrations of the drug results in cells whose proteins are hypomethylated [32; 33; 45]. Thus, pre-treating HeLa cells for 24 hr with 20 μM AdOx resulted in a substantial reduction in the recruitment of both FMRP and its homolog, FXR1P into *bona fide* stress granules, and an increase in a population of smaller cytoplasmic FMRP- and FXR1P-containing granules [14]. Furthermore, AdOx pre-treatment also resulted in the loss of FXR1P from arsenite-induced FMRP stress granules [14]. These data demonstrate that not all stress granules are equal and that their composition, which normally fluctuates, can be altered still more by the application of exogenous stimuli. This dynamism presumably has functional implications, related perhaps to sorting the various mRNAs within them into downstream pathways (triage or degradation). Future studies must address whether FMRP-containing stress granules are linked to one or both of these downstream pathways. Previous studies of stress granule formation, composition and dynamics have utilized HeLa cells [14; 38], COS-7 cells [28], HEK293 cells [8] and BHK cells [4], which have been referred to as "common cultured cell lines" [10]. However, as FMRP is thought to play a role in controlling local protein synthesis [63] it was of interest to determine whether FMRP-containing stress granules could form in a translational microdomain. To answer this question, we examined stress granule formation in PC12 cells grown in the presence of nerve growth factor (NGF) to induce neuritic extensions. Specifically, PC12 cells were first treated for 48 hr with NGF (200ng/ml). Subsequently, 0.5 mM sodium arsenite was added to the media and 0-20 min. later the cells were fixed and immunostained with anti-FMRP and anti-TIA-1 antibodies. The results of these studies showed that in response to sodium arsenite TIA-1 moved from the nucleus to the cytoplasm and into PC12 cell neurites and growth cones. More importantly, FMRP-containing stress granules were observed, not only in the soma, but were also present in both the neurites and the growth cones of these cells [15]. Thus, it would appear that all of the proteins necessary for stress granule formation are present in distal processes and can be recruited in response to an exogenous stressor. Therefore in response to this specific modulator, local and general protein synthesis appear to react similarly.

FMRP is a Component of Transport Granules

Kanai *et al* were among the first to characterize a *ca.* 1000S RNAse-sensitive granule isolated from mouse brain that bound to kinesin [25]. This granule was composed of 42 proteins and several dendritically targeted messenger RNAs, and similar granules were detected in cultured mouse hippocampal dendrites immunostained with various components of the isolated granule. Furthermore, these granules were shown to undergo microtubule-dependent bidirectional movement. Krichevsky *et al* who found similar granules in rat cerebral cortex primary neurons made two important observations concerning them [30]. First, they showed that these granules were translationally repressed as they could not support protein synthesis. This correlated with their lack of eukaryotic initiation factors 4E and 4G (eIF4E and eIF4G). Secondly, they showed that following KCl depolarization the mRNA in these granules re-localized to brain polyribosomes.

The existence of FMRP in brain transport granules, which was conclusively demonstrated by Kanai *et al* [25]were presaged by the studies of De Diego Otero *et al* in PC12 cell neurites[11]. Using inducible EGFP-FMRP heterologous expression, these authors showed that EGFP-FMRP granules, which harbored RNA, FXR1P and ribosomal subunits moved bi-directionally down neurites in a microtubule-dependent manner. Analogous results were provided by Antar *et al* using mouse hippocampal neurons and by Ling *et al* using the *Drosophila* homologue dFMR1 [2; 34]. These studies showed that FMRP, a negative regulator of translation [55] was an important component of transport granules.

Based on our prior work demonstrating that stress granules do not have a uniform composition, we inquired as to whether FMRP transport granules might also be heterogeneous as well. Hints that this was a real possibility existed in previous work. Specifically, elegant and detailed biochemical analyses by Mallardo *et al* had identified several types of staufen-containing complexes: large granules >2 MDa that contained ribosomal markers and two sub-populations of staufen RNA particles *ca.* 650 and 440 kDa, determined by the presence and absence of staufen-1 and staufen-2. The particles co-fractionated with microtuble markers, kinesin transport motor protein and contained CAMKIIα mRNA and BC-1 RNA, both of which co-localized with staufen in rat hippocampal dendrites, whereas the large granules lacked CAMKIIα mRNA and had substantially less BC-1 RNA [37]. Recent studies by Shina *et al* however, showed that the translational repressor RNG105 is found in large dendritic neuronal granules composed of ribosomal components as well as human staufen and CAMKIIα mRNA [50]. These data supported evidence of Kanai *et al* showing that a large neuronal transport granule contained staufen, CAMKIIα mRNA, a host of other RNA binding proteins (pur-α, pur-β, FMRP, FXR1P, FXR2P hnRNP U), and translation factors [25]. However, these data are not consistent with the findings of Antar *et al* demonstrating that FMRP and staufen-1 dendritic granules do not co-localize [2] or that FMRP and human staufen do not significantly co-localize [62], nor do they fit with the recent studies of Vanderklish *et al* showing that staufen-1 and FXR2P do not co-localize [61].

To address whether FMRP transport granules might be heterogeneous we took advantage of our prior work showing that arsenite treatment resulted in the recruitment of FMRP into stress granules and that stress granules could be generated in PC12 cell neurites [14; 15] by performing co-localization studies between FMRP and TIA-1, FMRP and FXR1P, and FMRP and FXR2P in differentiated PC12 cells that had been treated with sodium arsenite. Figure 3C shows typical results obtained for FMRP and TIA-1 co-localization. In the soma, there was near-complete co-localization between FMRP and TIA-1 following arsenite treatment (inset i); whereas, in the absence of arsenite the two proteins did not co-localize [15]. Similar results were found when we studied the granule distribution in the growth cones (inset ii); however, this was not the case when granules in the neurites were examined (inset iii). Here, the granules were clearly heterogeneous with *more than half* containing only FMRP. Comparable results were found when FMRP and FXR1P or FMRP and FXR2P were examined [15]. These results imply that complex and dynamic processes operate on the granules within PC12 cell neurites.

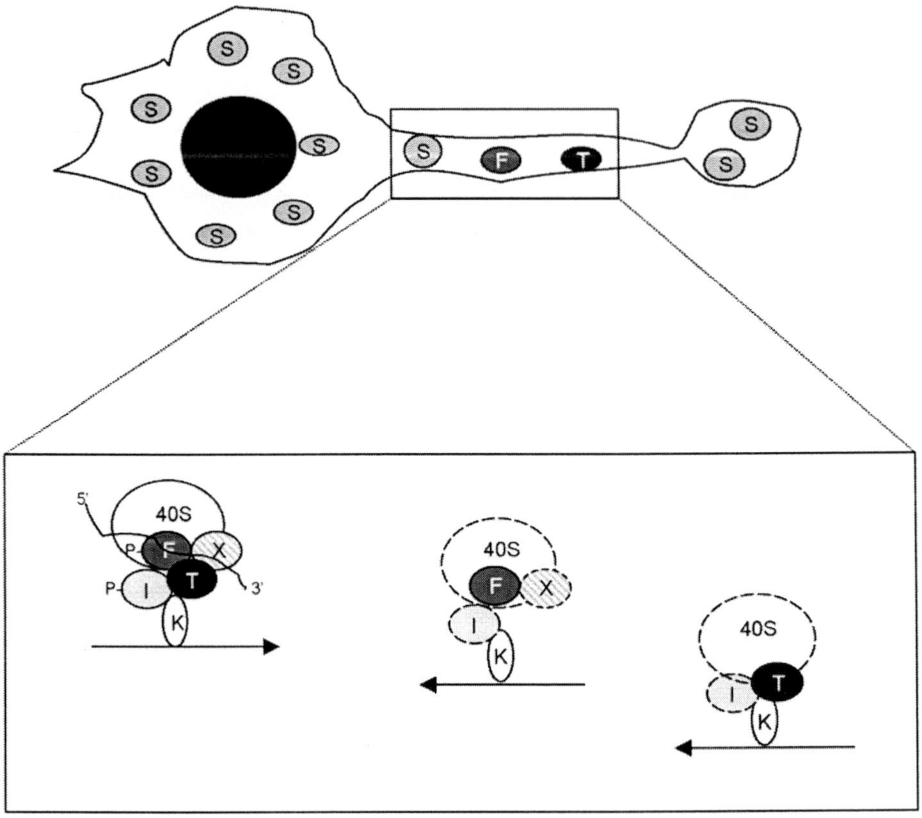

Figure 4. Model of FMRP trafficking in differentiated PC12 cells following arsenite treatment, based on immunofluorescence data in Figure 3B. Details are found in the text. Stress granules (S) and non-stress granules (F, or T) are shown as circles within the cell. A box, representing a magnified view of the granules is shown below the cell. Kinesin (K), FMRP (F), FXR1P or FXR2P (X), eIF-2α (I), TIA-1 (T) and 40 S subunits are labeled. P represents phosphorylated forms of eIF-2α and FMRP. Arrows depict the directionality of the microtubule network. Dashed circles indicate possible absence of a particular protein within a granule.

A model that is compatible with these data is shown in Figure 4. It is implicitly assumed that the formation of arsenite-induced stress granules marks messages that are potentially translationally competent. In the soma where stress granule formation first occurs most FMRP target messages are potentially translationally competent, although they may be repressed by association with the phosphorylated form of FMRP and/or with micro-RNA targets. Stress granule formation represses the synthesis of most messenger RNAs *via* the formation of abortive 48S* initiation complexes. The majority of these granules contain FMRP (some in the form of FMRP-FMRP homodimers, some in the form of FMRP-FXR1P or FMRP-FX2P heterodimers). As the local TIA-1 concentration increases in the cytoplasm through the export of TIA-1 from the nucleus, stress granules begin to form in the more distal regions of PC12 cells *i.e.* the neurites and growth cones marking other potentially translatable messages. Neuritic FMRP RNA transport granules are transported *via* microtubule motors and delivered to the growth cones. Prior to arsenite treatment, the granules that have transited down the neurites to the growth cones can deliver their cargo messages for local synthesis in response to NGF. Message delivery requires molecular rearrangement of the granules. Repressor proteins like FMRP and repressor RNAs must be jettisoned or inactivated, while essential initiation factors and 60S subunits must be recruited to form active polyribosomes. It is likely that plasticity-induced post-translational modifications to the transport granule proteins regulate this step, although the details of the process remain to be elucidated. The remnants of the transport granule are then shipped back to the soma through the neurites *via* microtubules. As these will lack messenger RNA and possibly other components they cannot form stress granules. Thus, arsenite treatment provides a snapshot of trafficking within neurites. We posit that the FMRP- and other stress granules in neurites mark messages enroute to growth cones, while the non-TIA-1-containing granules are protein and or protein-small RNA complexes that are transiting back to the nucleus to begin the formation of new transport granules.

While the broad outlines of this model are compatible with other models of dendritic transport and delivery [2; 5; 22; 41; 61; 68] future studies must provide the complete mechanistic details. In particular, the formation of FMRP-containing transport granules and their recruitment to the microtuble transport system are poorly understood. Similarly, the molecular basis of their activity-dependent delivery remains to be addressed.

Conclusion

From this rudimentary analysis it is clear that the RNA binding protein FMRP leads a complex life as a component of several different RNA granules. This life, whose details are increasingly becoming known, encompasses much of what would be referred to as the biology of RNA. Nevertheless, our current picture of FMRP RNA granules is probably incomplete for two reasons. First, the distinctions between RNA granule types are often blurred. For instance, are all mRNA transport granules identical? If not, what are the functional ramifications of differences in composition due to the mRNA load, the protein components or the type of repressor and how can individual granules be studied in detail? Second, as mentioned in the introduction, it is likely that there are other FMRP granules than the ones focused on in this chapter. Indeed, a preliminary study from the Greenough laboratory characterizing a novel FMRP RNA granule has recently been presented [22]. Thus, it is likely that the study of RNA granules will be a fruitful area of investigation for many years.

Acknowledgement

We would like to thank Dr. George Merz for help in obtaining the confocal images presented here. Finally, thanks to Dr. David L Miller for his helpful comments on this manuscript. This work was supported by the New York State Research Foundation for Mental Hygiene.

REFERENCES

[1] Anderson, P. and Kedersha, N. (2002). Stressful initiations. *J Cell Sci* 115, 3227-3234.

[2] Antar, L. N., Dictenberg, J. B., Plociniak, M., Afroz, R. and Bassell, G. J. (2005). Localization of FMRP-associated mRNA granules and requirement of microtubules for activity-dependent trafficking in hippocampal neurons. *Genes Brain Behavior* 4, 350-359.

[3] Ashley, C. T., Jr., Wilkinson, K. D., Reines, D. and Warren, S. T. (1993). FMR1 protein: conserved RNP family domains and selective RNA binding. *Science* 262, 563-6.

[4] Baez, M. V. and Boccaccio, G. L. (2005). Mammalian smaug is a translational repressor that forms cytoplasmic foci similar to stress granules. *J. Biol. Chem.*, M508374200.

[5] Bassell, G. J. and Twiss, J. (2006). RNA exodus to Israel: RNA controlling function in the far reaches of the neuron. *EMBO Rep* 7, 31-35.

[6] Cao, Q. and Richter, J. D. (2002). Dissolution of the maskin-eIF4E complex by cytoplasmic polyadenylation and poly(A)-binding protein controls cyclin B1 mRNA translation and oocyte maturation. *EMBO J* 21, 3852-3862.

[7] Caudy, A. A., Myers, M., Hannon, G. J. and Hammond, S. M. (2002). Fragile X-related protein and VIG associate with the RNA interference machinery. *Genes Dev* 16, 2491-6.

[8] Cougot, N., Babajko, S. and Seraphin, B. (2004). Cytoplasmic foci are sites of mRNA decay in human cells. *J. Cell Biol.* 165, 31-40.

[9] Crino, P. B. and Eberwine, J. (1996). Molecular Characterization of the Dendritic Growth Cone: Regulated mRNA Transport and Local Protein Synthesis. *Neuron* 17, 1173-1187.

[10] Davidovic, L., Tremblay, S., De Koninck, P. and Khandjian, E. (2005). Fragile X mental retardation protein: To be or not to be a translational repressor. In *The Molecular Basis of Fragile X Syndrome*, (eds. Y-J. Sung and R. B. Denman), pp. 129-143. Trivandrum, India: Research Signpost.

[11] De Diego Otero, Y., Severijnen, L-A., van Cappellen, G., Schrier, M., Oostra, B. and Willemsen, R. (2002). Transport of fragile X mental retardation protein via granules in neurites of PC12 Cells. *Mol. Cell. Biol.* 22, 8332-8341.

[12] Denman, R. B. (2003). Deja vu all over again: FMRP binds U-rich target mRNAs. *Biochemical and Biophysical Research Communications* 310, 1-7.

[13] Denman, R. B., Dolzhanskaya, N. and Merz, G. (2004). Effects of PRMT methylation on FMRP expression, localization and function. In *Society for Neurosciences Annual Meeting*, #239.2. San Diego, CA.

[14] Dolzhanskaya, N., Merz, G., Aletta, J. M. and Denman, R. B. (2006). Methylation regulates FMRP's intracellular protein-protein and protein-RNA interactions. *J Cell Sci*, (in press).

[15] Dolzhanskaya, N., Merz, G. and Denman, R. B. (2005). Identifying translationally-competent endogenous FMRP granules in PC12 Cell neurites by oxidative stress. *Molecular Brain Research* (submitted).

[16] Froster, U. G. (1995). Fragile X syndrome: clinical and molecular genetics correlations. *Ther Umsch* 52, 805-9.

[17] Gibson, T. J., Rice, P. M., Thompson, J. D. and Heringa, J. (1993). KH domains within the FMR1 sequence suggest that fragile X syndrome stems from a defect in RNA metabolism. *Trends Biochem Sci* 18, 331-3.

[18] Gilks, N., Kedersha, N., Ayodele, M., Shen, L., Stoecklin, G., Dember, L. M. and Anderson, P. (2004). Stress granule assembly is mediated by prion-like aggregation of TIA-1. *Mol. Biol. Cell* 15, 5383-5398.

[19] Grigston, J. C., VanDongen, H. M. A., McNamara II, J. O. and VanDongen, A. M. J. (2005). Translation of an integral membrane protein in distal dendrites of hippocampal neurons. *European Journal of Neuroscience* 21, 1457-1468.

[20] Grudzien, E., Kalek, M., Jemielity, J., Darzynkiewicz, E. and Rhoads, R. E. (2005). Differential inhibition of mRNA degradation pathways by novel cap analogs. *J. Biol. Chem.*, M509121200.

[21] Hagerman, R. J. (1996). Physical and behavioral phenotype. In *Fragile X Syndrome: Diagnosis, Treatment and Research,* (eds. R. J. Hagerman, Cronister, A.), pp. 3-87. Baltimore, MD: Johns Hopkins University Press.

[22] Hou, X., Weiler, I. J. and Greenough, W. T. (2005). A novel heavy FMRP granule not containing ribosomes or kinesin. In *Society for Neurosciences Annual Meeting,* # 115.18. Washington, D.C.

[23] Hua, Y. and Zhou, J. (2004). Survival motor neuron protein facilitates assembly of stress granules. *FEBS Lett* 572, 617-622.

[24] Huang, Y-S. and Richter, J. D. (2004). Regulation of local mRNA translation. *Current Opinion in Cell Biology* 16, 308-313.

[25] Kanai, Y., Dohmae, N. and Hirokawa, N. (2004). Kinesin Transports RNA: Isolation and Characterization of an RNA-Transporting Granule. *Neuron* 43, 513-525.

[26] Kayali, F., Montie, H. L., Rafols, J. A. and DeGracia, D. J. (2005). Prolonged translation arrest in reperfused hippocampal cornu Ammonis 1 is mediated by stress granules. *Neuroscience* 134, 1223-1245.

[27] Kedersha, N., Chen, S., Gilks, N., Li, W., Miller, I. J., Stahl, J. and Anderson, P. (2002). Evidence That Ternary Complex (eIF2-GTP-tRNAiMet)-Deficient Preinitiation Complexes Are Core Constituents of Mammalian Stress Granules. *Mol. Biol. Cell* 13, 195-210.

[28] Kedersha, N., Stoecklin, G., Ayodele, M., Yacono, P., Lykke-Andersen, J., Fitzler, M. J., Scheuner, D., Kaufman, R. J., Golan, D. E. and Anderson, P. (2005). Stress granules and processing bodies are dynamically linked sites of mRNP remodeling. *J. Cell Biol.* 169, 871-884.

[29] Kim, S., Weiler, I. J. and Greenough, W. T. (2005). Translocation of mRNAs by FMRP between stress granules and polyribosomes after synaptic stimulation. In *Society for Neuroscience Annual Meeting*, #115.17. Washington, D.C.

[30] Krichevsky, A. M. and Kosik, K. S. (2001). Neuronal RNA Granules: A Link between RNA Localization and Stimulation-Dependent Translation. *Neuron* 32, 683-696.

[31] Kuersten, S. and Goodwin, E. B. (2003). The power of the 3' UTR: translational control and development. *Nature Reviews Genetics* 4, 626-637.

[32] Li, C., Ai, L. S., Lin, C. H., Hsieh, M., Li, Y. C. and Li, S. Y. (1998). Protein N-arginine methylation in adenosine dialdehyde-treated lymphoblastoid cells. *Arch Biochem Biophys* 351, 53-9.

[33] Lin, C. H., Hsieh, M., Li, Y. C., Li, S. Y., Pearson, D. L., Pollard, K. M. and Li, C. (2000). Protein N-arginine methylation in subcellular fractions of lymphoblastoid cells. *J Biochem (Tokyo)* 128, 493-8.

[34] Ling, S-C., Fahrner, P. S., Greenough, W. T. and Gelfand, V. I. (2004). Transport of Drosophila fragile X mental retardation protein-containing ribonucleoprotein granules by kinesin-1 and cytoplasmic dynein. *Proc Nat'l Acad Sci* 101, 17428-17433.

[35] Liu, J., Valencia-Sanchez, M. A., Hannon, G. J. and Parker, R. (2005). MicroRNA-dependent localization of targeted mRNAs to mammalian P-bodies. *Nat Cell Biol* 7, 719-723.

[36] Lugli, G., Larson, J., Martone, M. E., Jones, Y. and Smalheiser, N. R. (2005). Dicer and eIF2c are enriched at postsynaptic densities in adult mouse brain and are modified by neuronal activity in a calpain-dependent manner. *Journal of Neurochemistry* 94, 896-905.

[37] Mallardo, M., Deitinghoff, A., Muller, J., Goetze, B., Macchi, P., Peters, C. and Kiebler, M. A. (2003). Isolation and characterization of Staufen-containing ribonucleoprotein particles from rat brain. *PNAS* 100, 2100-2105.

[38] Mazroui, R., Huot, M-E., Tremblay, S., Filion, C., Labelle, Y. and Khandjian, E. W. (2002). Trapping of messenger RNA by Fragile X Mental Retardation protein into cytoplasmic granules induces translation repression. *Hum. Mol. Genet.* 11, 3007-3017.

[39] McDunn, J. E. and Cobb, J. P. (2005). That Which Does Not Kill You Makes You Stronger: A Molecular Mechanism for Preconditioning. *Sci. STKE* 2005, pe34-.

[40] Mendez, R. and Richter, J. D. (2001). Translational control by CPEB: a means to the end. *Nat Rev Mol Cell Biol* 2, 521-529.

[41] Miki, T., Takano, K. and Yoneda, Y. (2005). The role of mammalian staufen on mRNA traffic: a view from its nucleoplasmic shuttling function. *Cell Struct Funct* 30, 51--56.

[42] Moore, M. J. (2005). From Birth to Death: The Complex Lives of Eukaryotic mRNAs. *Science* 309, 1514-1518.

[43] Muslimov, I. A., Banker, G., Brosius, J. and Tiedge, H. (1998). Activity-dependent Regulation of Dendritic BC1 RNA in Hippocampal Neurons in Culture. *J. Cell Biol.* 141, 1601-1611.

[44] Musumeci, S. A., Hagerman, R.J., Ferri, R.Bosco, P., Dalla Bernardina, B., Tassinari, C. A., DeSarro, G. B., and Elia, M. (1999). Epilepsy and EEG findings in males with fragile X syndrome. *Epilepsia* 40, 1092-9.

[45] Nichols, R. C., Wang, X. W., Tang, J., Hamilton, B. J., High, F. A., Herschman, H. R. and Rigby, W. F. (2000). The RGG domain in hnRNP A2 affects subcellular localization. *Exp Cell Res* 256, 522-32.

[46] Ogawa, F., Kasai, M. and Akiyama, T. (2005). A functional link between Disrupted-In-Schizophrenia 1 and the eukaryotic translation initiation factor 3. *Biochemical and Biophysical Research Communications* 338, 771-776.

[47] Pham, J. W. and Sontheimer, E. J. (2005). Molecular Requirements for RNA-induced Silencing Complex Assembly in the Drosophila RNA Interference Pathway. *J. Biol. Chem.* 280, 39278-39283.

[48] Reiss, A. L., Freund, L. S., Baumgardner, T. L., Abrams, M. T. and Denckla, M. B. (1995). Contribution of the FMR1 gene mutation to human intellectual dysfunction. *Nat Genet* 11, 331-4.

[49] Shav-Tal, Y. and Singer, R. H. (2005). RNA localization. *J Cell Sci* 118, 4077-4081.

[50] Shiina, N., Shinkura, K. and Tokunaga, M. (2005). A novel RNA-binding protein in neuronal RNA granules: regulatory machinery for local translation. *J. Neurosci.* 25, 4420-4434.

[51] Singer, R. H. (2003). RNA localization: visualization in real-time. *Current Biology* 13, R673-R675.

[52] Siomi, H., Siomi, M. C., Nussbaum, R. L. and Dreyfuss, G. (1993). The protein product of the fragile X gene, FMR1, has characteristics of an RNA-binding protein. *Cell* 74, 291-8.

[53] Sontheimer, E. J. (2005). Assembly and function of RNA silencing complexes *Nature Reviews Molecular Cell Biology* 6, 127-138.

[54] Steward, O. and Schuman, E. M. (2003). Compartmentalized Synthesis and Degradation of Proteins in Neurons. *Neuron* 40, 347-359.

[55] Sung, Y-J., Dolzhanskaya, N., Nolin, S. L., Brown, W. T., Currie, J. R. and Denman, R. B. (2003). The fragile X mental retardation protein FMRP binds elongation factor 1A mRNA and negatively regulates its translation in vivo. *J. Biol. Chem.* 278, 15669-15678.

[56] Sung, Y-J., Walters, E. T. and Ambron, R. T. (2004). A neuronal isoform of protein kinase G couples mitogen-activated protein kinase nuclear import to axotomy-induced long-term hyperexcitability in Aplysia sensory neurons. *J. Neurosci.* 24, 7583-7595.

[57] Sutton, M. A. and Schuman, E. M. (2005). Local translational control in dendrites and its role in long-term synaptic plasticity. *Journal of Neurobiology* 64, 116-131.

[58] Taddei, A., Hediger, F., Neumann, F. R., Bauer, C. and Gasser, S. M. (2004). MK2-induced tristetraprolin: 14-3-3 complexes prevent stress granule association and ARE-mRNA decay. *EMBO J* 23, 1301–1312.

[59] Teixeira, D., Sheth, U., Valencia-Sanchez, M. A., Brengues, M. and Parker, R. (2005). Processing bodies require RNA for assembly and contain nontranslating mRNAs. *RNA* 11, 371-382.

[60] Tourriere, H., Chebli, K., Zekri, L., Courselaud, B., Blanchard, J. M., Bertrand, E. and Tazi, J. (2003). The rasGAP-associated endoribonuclease G3BP assembles stress granules. *J Cell Biol* 160, 823-831.

[61] Vanderklish, P. W. and Edelman, G. M. (2005). Differential translation and fragile X syndrome. *Genes Brain Behavior* 4, 360-384.

[62] Villace, P., Marion, R. M. and Ortin, J. (2004). The composition of staufen-containing RNA granules from human cells indicates their role in the regulated transport and translation of messenger RNAs. *Nucl. Acids. Res.* 32, 2411-2420.

[63] Weiler, I-J. (2005). FMRP and the regulation of protein translation near synapses. In *The Molecular Basis of Fragile X Syndrome*, (eds Y-J. Sung and R. B. Denman), pp. 201-215. Trivandrum, India: Research Signpost.

[64] Weiler, I-J., Irwin, S. A., Klintsova, A. Y., Spencer, C. M., Brazelton, A. D., Miyashiro, K., Comery, T. A., Patel, B., Eberwine, J. and Greenough, W. T. (1997). Fragile X mental retardation protein is translated near synapses in response to neurotransmitter activation. *Proc Natl Acad Sci U S A* 94, 5395-400.

[65] Wilczynska, A., Aigueperse, C., Kress, M., Dautry, F. and Weil, D. (2005). The translational regulator CPEB1 provides a link between dcp1 bodies and stress granules. *J Cell Sci* 118, 981-992.

[66] Yu, J. H., Yang, W.-H., Gulick, T. O. D., Bloch, K. D. and Bloch, D. B. (2005). Ge-1 is a central component of the mammalian cytoplasmic mRNA processing body. *RNA* 11, 1795-1802.

[67] Zhang, L., King, O., Wong, S., Goldberg, D., Tong, A., Lesage, G., Andrews, B., Bussey, H., Boone, C. and Roth, F. (2005). Motifs, themes and thematic maps of an integrated Saccharomyces cerevisiae interaction network. *Journal of Biology* 4, 6.

[68] Zhong, J. and Tiedge, H. (2005). Role of FMRP in translational control. In *The Molecular Basis of Fragile X Syndrome*, (eds. Y-J. Sung and R. B. Denman), pp. 217-228. Trivandrum, India: Research Signpost.

INDEX

A

B

C

D

E

F

G

H

I

J

K

L

M

N

R

S

T

U

V